- Biography Series -

SCIENTISTS OF INTERNATIONAL REPUTE

By
Shanti Gopalan

SURA BOOKS (Pvt) LTD.
Chennai ● Bangalore ● Kolkata ● Ernakulam

Price: Rs.40.00

© PUBLISHERS

SCIENTISTS OF INTERNATIONAL REPUTE

By
Shanti Gopalan

First Edition : January, 2005

Second Print : January, 2006

Size : ⅛ Demy

Pages : 96

Price: Rs.40.00

ISBN: 81-7478-628-7

[NO ONE IS PERMITTED TO COPY OR TRANSLATE IN ANY OTHER LANGUAGE THE CONTENTS OF THIS BOOK OR PART THEREOF IN ANY FORM WITHOUT THE WRITTEN PERMISSION OF THE PUBLISHERS]

SURA BOOKS (PVT) LTD.

Head Office:
1620, 'J' Block,
16th Main Road,
Anna Nagar,
Chennai - 600040.
Phones: 26162173, 26161099.

Branch:
XXXII/2328, New Kalavath Road,
Opp. to BSNL, Near Chennoth Glass,
Palarivattom,
Ernakulam - 682025.
Phone: 0484-3205797

Printed at T. Krishna Press, Chennai - 600 102 and
Published by V.V.K.Subburaj for Sura Books (Pvt) Ltd.,
1620, 'J' Block, 16th Main Road, Anna Nagar, Chennai - 600 040.
Phones: 26162173, 26161099. Fax: (91) 44-26162173.
email: surabooks@eth.net; website: www.surabooks.com

CONTENTS

Page

ASTRONOMY 1
    STEPHEN HAWKING 4

MATHEMATICS 14
    SRINIVASA RAMANUJAM 17

ASTRONAUT 26
    KALPANA CHAWLA 26

AGRICULTURE 39
    M.S. SWAMINATHAN 39

THE NOBEL LAUREATES 45
    SIR C.V. RAMAN 45
    SUBRAHMANYAN CHANDRASEKHAR 66

ORNITHOLOGIST 85
    Dr. SALIM ALI 85

BIOGRAPHY SERIES

SCIENTISTS OF INTERNATIONAL REPUTE

ASTRONOMY

Today's science is tomorrow's solution. We are now standing in the threshold of the twenty-first century, advancing in our knowledge about the secrets of our world as never before. We live in an exploring age. In search of treasure and discovery we go down to the floor of the sea, journey towards the stars. With the advancement of science and technology we can comprehend the quasars and pulsars in the distant heaven.

Man's most magnificent achievement and a monumental effort is the real understanding of our universe. There are many scientists and explorers of today who deserve honour from all of us for the incredible revelation of the riddle called the 'universe'! If not for them, even after these hundreds of thousands of years later, universe being as remote from us as we are from the time the world started, we will be still looking at it with the same kind of wonder and ignorance as we did as uninformed humans! These were the torch-bearers who made it clear that human kind clings to but one thread of the web of life in this vast universe!

All your life, you have been fascinated by the big questions that face us, and have tried to find scientific answers to them. Where do we come from? How did the universe begin? Why is the universe the way it is? How will it end? Have you been intrigued by the questions about the stars or about their origin? Have you searched for answers for some of our deepest questions like how did our Universe originate? How old it is? Is there life in Mars? Have you a feeling that life exists outside the earth? Did extra-terrestrial life exist much before we came to the world? How do we go about answering such questions?

There are no perfect, proven answers because on earth humans have already erased the clues for the origins of the earth. But on Mars we have found rocks older than the earthen rocks and Mars is fresh with the features of its origin as it has been left undisturbed. Did you know that life may have existed- still exist outside the earth which might even be more advanced than our own race. Recently they have discovered a planet which is at the same distance from its star as the earth is from the sun. So scientists feel that life may exist in that far away planet as the heat there is almost the same as the heat felt on earth. 'Europa' a satellite of Jupiter, in our solar system has water on its surface and missions are planned to explore this heavenly body. Some of the clues for extra-terrestrial life are water, oxygen and carbon which are the main ingredients for life. How is a 'black hole' formed ? If you have looked at the stars, and tried to make sense of what you see, you too have started to wonder what makes the universe exist.

Night after night the stars appear to be standing still, but it is an illusion caused by the distance. Actually it will be hurtling

ASTRONOMY

through the space at more than 1,50,000 kms an hour! The ideas which had grown over two thousand years of observation are to be radically revised. In less than a hundred years, we have found a new way to think of ourselves. From sitting at the center of the universe, we now find ourselves orbiting an average-sized sun, which is just one of the millions of stars in our own Milky Way galaxy. And our galaxy itself is just one of billions of galaxies, in a universe that is infinite and expanding. But this is far from the end of a long history of inquiry. Many questions remain to be answered, before we can hope to have a complete picture of the universe we live in. Mars is about 5 billion years old and the universe originated 15 billion years ago. Based on the present studies life has not been found on Mars except for the fossils of bacteria on the meteorite ALH 84001 which landed on the earth and is believed to have come from Mars.

A Black Hole is formed after the death of a star. It is very dense that its gravitation bends the light rays that inclines towards it and those closer to the black hole are even attracted into it . In recent years we have found light bending and coming to us from distant stars and thus we have detected proof of Black Holes.

The questions about our universe are clear, and deceptively simple, but the answers have always seemed well beyond our reach, until the turn of the century, when men and women committed themselves to unraveling the mystery of the universe. This book represents the study of this human physical/intellectual effort of the highest order that went into the study of the universe. It talks about the men/women behind the real excavation of our physical universe, one of man's most magnificent achievements. Just as we try to understand the world we live in with the help of these scientists, stimulating the search for ways to live in Mars by future young scientists could produce tremendous rewards for mankind. Today we have an idea where we came from, but not where we are heading as species. If we could 'live' on Mars, we could be sure that we are a part of the universal scheme of things, and a measure of how far human mind and spirit could evolve. This book tries to inspire the motivation of our young, future scientists.

STEPHEN HAWKING

Stephen William Hawking is one of the world's leading theoretical physicists. He has been compared to Isaac Newton by some observers. His principal areas of research are cosmology and quantum gravity. One of his major contributions to the field of research were his papers on the relationship between black holes and thermodynamics. His research indicated that black holes do not exist forever, but rather that virtual particle pairs created near their event horizons cause them to "evaporate" over time.

Stephen William Hawking was born on 8 January 1942 (300 years after the death of Galileo) in Oxford, England. His parents' house was in north London, but during the second world war Oxford was considered a safer place to have babies. Stephen Hawking's parents lived in London where his father was undertaking research into medicine. However, London was a dangerous place during World War II and Stephen's mother was sent to the safer town of Oxford where Stephen was born. The family was soon back together living in Highgate, north London, where Stephen began his schooling. When he was eight, his family moved to St. Albans, a town about 20 miles north of London.

At eleven Stephen went to St. Albans School, and then on to University College, Oxford, where his father studied. Stephen wanted to do Mathematics, although his father would have preferred medicine. Mathematics was not available at University College, so he did Physics instead. After three years and not very much work he was awarded a first class honours degree in Natural Science.

In 1950 Stephen's father moved to the Institute for Medical Research in Mill Hill. The family moved to St. Albans so that the journey to Mill Hill was easier. Stephen attended St. Albans High School for Girls (which took boys up to the age of 10). His father wanted him to take the scholarship examination to go to Westminster Public School. However Stephen was ill at the time of the examinations and remained at St. Albans school which he had attended from the age of 11.

Stephen writes: *I got an education there that was as good as, if not better than, that I would have had at Westminster. I have never found that my lack of social graces has been a hindrance."*

Hawking wanted to specialise in mathematics in his last couple of years at school where his mathematics teacher had inspired him to study the subject. However Hawking's father was strongly against the idea and Hawking was persuaded to make chemistry his main school subject. Part of his father's reasoning was that he wanted Hawking to go to University College, Oxford, the College he himself had attended, and that College had no mathematics fellow.

In March 1959, Hawking took the scholarship examinations with the aim of studying natural sciences at Oxford. He was awarded a scholarship, and at University College he specialised in physics in his natural sciences degree. He only just made a First Class degree in 1962.

The prevailing attitude at Oxford at that time was very anti-work. You were supposed to be brilliant without effort, or accept your limitations and get a fourth-class degree. To work hard to get a better class of degree was regarded as the mark of a grey man - the worst epithet in the Oxford vocabulary.

From Oxford, Hawking moved to Cambridge to take up research in general relativity and cosmology, a difficult area for someone with only a little mathematical background. Hawking had noticed that he was becoming rather clumsy during his last year at Oxford and, when he returned home for Christmas in 1962 at the end of his first term at Cambridge, his mother persuaded him to see a doctor.

In early 1963 he spent two weeks undergoing tests in hospital and motor neurone disease (Lou Gehrig's disease) was diagnosed. His condition deteriorated quickly and the doctors predicted that he would not live long enough to complete his doctorate.

However Hawking writes: *"Although there was a cloud hanging over my future, I found to my surprise that I was enjoying life in the present more than I had before. I began to make progress with my research."*

The reason that his research progressed was that he met a girl he wanted to marry and realised he had to complete his doctorate to get a job.

Hawking writes:- *"I therefore started working for the first time in my life. To my surprise I found I liked it."*

After completing his doctorate in 1966 Hawking was awarded a fellowship at Gonville and Caius College, Cambridge.

Stephen then went on to Cambridge to do research in Cosmology, there being no-one working in that area in Oxford at that time. His supervisor was Denis Sciama, although he had hoped to get Fred Hoyle who was working in Cambridge. At first his position was that of Research Fellow, but after gaining his Ph.D. he became first, a Research Fellow, and later on a

Professorial Fellow at Gonville and Caius College. After leaving the Institute of Astronomy in 1973 Stephen came to the Department of Applied Mathematics and Theoretical Physics, and since 1979 has held the post of Lucasian Professor of Mathematics. The chair was founded in 1663 with money left by the will of the Rev. Henry Lucas, who had been the Member of Parliament for the University. The chair was first held by Isaac Barrow, and then in 1663 by Isaac Newton. The man born 300 years after Galileo died now held Newton's chair at Cambridge.

He became Professor of Gravitational Physics at Cambridge in 1977. In 1979 Hawking was appointed Lucasian Professor of Mathematics at Cambridge.

Between 1965 and 1970 Hawking worked on singularities in the theory of general relativity devising new mathematical techniques to study this area of cosmology. Much of his work in this area was done in collaboration with Roger Penrose who, at that time, was at Birkbeck College, London. From 1970 Hawking began to apply his previous ideas to the study of black holes. Continuing this work on black holes, Hawking discovered in 1970 a remarkable property. Using quantum theory and general relativity he was able to show that black holes can emit radiation. His success at proving this made him work from that time on combining the theory of general relativity with quantum theory. In 1971 Hawking investigated the creation of the Universe and predicted that, following the big bang, many objects as heavy as 10^9 tons but only the size of a proton would be created. These mini black holes have large gravitational attraction governed by general relativity, while the laws of quantum mechanics would apply to objects that are small.

Another remarkable achievement of Hawking's using these techniques was his 'no boundary proposal' made in 1983 with Jim Hartle of Santa Barbara. Hawking explains that this would mean:-

"That both time and space are finite in extent, but they don't have any boundary or edge. There would be no singularities, and the laws of science would hold everywhere, including at the beginning of the universe."

In 1982 Hawking decided to write a popular book on cosmology. By 1984 he had produced a first draft of "A Brief History of Time." However Hawking was to suffer a further illness:

"I was in Geneva, at CERN, the big particle accelerator, in the summer of 1985. I caught pneumonia and was rushed to hospital. The hospital in Geneva suggested to my wife that it was not worth keeping the life support machine on. But she was having none of that. I was flown back to Addenbrooke's Hospital in Cambridge, where a surgeon called Roger Grey carried out a tracheotomy. That operation saved my life but took away my voice."

Hawking was given a computer system to enable him to have an electronic voice. It was with these difficulties that he revised the draft of 'A Brief History of Time' which was published in 1988. The book broke sales records in a way that it would have been hard to predict. By May 1995 it had been in The Sunday Times best-sellers list for 237 weeks breaking the previous record of 184 weeks. This feat is recorded in the 1998 Guinness Book of Records. Also recorded there is the fact that the paperback edition was published on 6 April 1995 and reached number one in the best sellers in 3 days. By April 1993 there had been 40 hardback editions of 'A Brief History of Time' in the United States and 39 hardback editions in the UK.

Of course Hawking has received, and continues to receive, a large number of honours. He was elected a Fellow of The Royal Society in 1974, being one of its youngest fellows. He was awarded the CBE in 1982, and was made a Companion of Honour in 1989. Hawking has also received many foreign awards and prizes and was elected a Member of the National Academy of Sciences of the United States.

Stephen Hawking has worked on the basic laws which govern the universe. With Roger Penrose he showed that Einstein's General Theory of Relativity implied space and time, and would have a beginning in the Big Bang and an end in black holes. These results indicated that it was necessary to unify General Relativity with Quantum Theory, the other great Scientific development of the first half of the 20th Century. One consequence of such a unification that he discovered was that black holes should not be completely black, but should emit radiation and eventually evaporate and disappear. Another conjecture is that the universe has no edge or boundary in imaginary time. This would imply that the way the universe began was completely determined by the laws of science.

In the late 1960s, Hawking proved that if general relativity is true and the universe is expanding, a singularity must have occurred at the birth of the universe. In 1974 he first recognized a truly remarkable property of black holes, objects from which nothing was supposed to be able to escape. By taking into account quantum mechanics, he was able to show that black holes can radiate energy as particles are created in their vicinity.

His many publications include The Large Scale Structure of Space Time with G.F.R. Ellis, General Relativity: An Einstein Centenary Survey, with W Israel, and 300 Years of Gravity, with W Israel. Stephen Hawking has two popular books published; his best seller A Brief History of Time, and his later book, Black Holes and Baby Universes and Other Essays.

Stephen Hawking has devoted much of his life to probing the space-time described by general relativity and the singularities where it breaks down. And he has done most of this work while confined to a wheelchair, brought on by the progressive neurological disease amyotrophic lateral sclerosis, or Lou Gehrig's Disease, but he never felt trapped by the situation.. Hawking is still the Lucasian Professor of Mathematics at Cambridge, a post once held by Isaac Newton.

Stephen Hawking continues to combine family life (he has three children and one grandchild), and his research into theoretical physics together with an extensive programme of travel and public lectures.

Stephen Hawking's genius is universally acknowledged now. Time will prove that he was right on every count. His findings inspired a greater awe, then avid curiosity, or at least, a wider speculation. The feeling that his theories are the most intriguing possibilities is deep and intense. For the sheer creative energy brought about by a single individual, at a time when we are wearily getting used to the idea that there is nothing more single individual can do in the turn of this Century, even in this age of giants, he walks tall.

Stephen Hawking has worked on the basic laws which govern the universe. With Roger Penrose he showed that Einstein's General Theory of Relativity implied space and time would have a beginning in the Big Bang and an end in black holes. These results indicated it was necessary to unify General Relativity with Quantum Theory, the other great Scientific development of the first half of the 20th Century. One consequence of such a unification that he discovered was that black holes should not be completely black, but should emit radiation and eventually evaporate and disappear. Another conjecture is that the universe has no edge or boundary in imaginary time. This would imply that the way the universe began was completely determined by the laws of science.

The massive star starts to collapse when it exhausts its nuclear fuel and can no longer counteract the inward pull of gravity. The crushing weight of the star's overlying layers implodes the core, and the star digs deeper into the fabric of space-time. Although the star remains barely visible, its light now has a difficult time climbing out of the enormous gravity of the still-collapsing core. The star passes through its event horizon and disappears from our universe, forming a singularity of infinite density. Pile enough matter into a small enough volume and its gravitational pull will grow so strong that nothing can escape from it. That includes light, which travels at the absolute cosmic speed limit of 186,000 miles per second. In a stroke of descriptive genius, physicist John Wheeler named these objects "black holes." The radius of a black hole is called the event horizon because it marks the edge beyond which light cannot escape, so any event taking place inside the event horizon can never be glimpsed from outside—in effect, the inside of the black hole is cut off from our universe. It has even been speculated that black holes could be pathways into other universes. Gravity is so strong at the center of a black hole, that even Einstein's gravitational laws must break down. The theory that governs the incredibly dense matter and strong gravitational fields at the center of a black hole is not yet known." He wanted the world to share his excitement of his discoveries, both past and present, which has revolutionized our way of thinking.

Hawking celebrated his 60th birthday recently, quite an achievement for someone given a couple of years to live nearly 40 years ago. His two books 'A Brief History of Time' and 'The Universe in a Nutshell' have remained highly popular all over the world and are now classic best-sellers. Anyone interested in universe, cosmos and how it all began can read them: no previous knowledge in this field is required to enjoy these books.

He had a guest appearance on an episode of the television series Star Trek: The Next Generation, playing poker with Data, Albert Einstein, and Isaac Newton in the episode "Descent, Part I".

The animated television series The Simpsons has occasionally featured him in episodes.

In November 2003 Intel donated a new computer to Professor Hawking, custom-designed for his wheelchair. The computer is a modified laptop running on a 1.5 GHz Pentium M chip, with Centreno technology.

From the following report on the activities of the professor, we can see how active he is in spite of his disability. In January 2003 Professor Hawking gave a lecture discussing the future of physics, for the Cambridge-MIT Institute. The lecture was transmitted live to MIT in America via the web. In February 2003 Professor Hawking spent a month attending the Holography conference at Texas A & M University. During that time he lectured in College Station, Galveston and the Woodlands. In March 2003 Professor Hawking spent a week at the Cosmic Inflation conference at UC Davis in California. In June 2003 Professor Hawking made a surprise appearance on the 'Conan O'Brian Show' in America, along side Jim Carrey. A week later Jim dropped in to Cambridge to join him for dinner. In August 2003 Professor Hawking attended the Nobel Symposium on String Theory and Cosmology, in Sigtuna, Sweden. In September 2003 Professor Hawking spent 7 weeks in the USA, visiting Caltech in Los Angeles, and attending the KITP conference at UCSB in Santa Barbara. He also spent a few days in Cleveland to take part in the CERCA conference at CWRU, and gave a public lecture in Severance Hall. In October 2002 Professor Hawking travelled to Paris to give a lecture on Brane theory at the Evergreen Partners Annual Meeting at the Four Seasons Hotel.

In September 2002 Professor Hawking received a new computer system from Intel. This new converted laptop runs at over 1 GHz and has a significantly different appearance to his last system. Professor Hawking received acclaims and prizes for his new book, 'The Universe in a Nutshell'. The book is a compilation of seven of the most ground-breaking books in the history of science, together with five critical essays and a biography of each featured physicist, written by Hawking himself. Professor Hawking said that though he did not expect to win prizes, he was very much pleased in it. He said that Science writing really can have an impact on how we live. He found the thirst in people to know more wherever he goes all around the world. And that this 'want' has helped raise the profile of science.

A new IMax movie by Stephen Hawking's 'Beyond the Horizon' was expected to be released by late 2004. It enlightens the lives of the earliest cave dwellers, to the Babylonians and the Greeks, the world's great temples, art, and literature, and is sought to bring the public closer to the cosmos. It is impossible to date the beginning of humankind's fascination with the stars and the world beyond them. Now, the world's most important large-format filmmakers, scientists, and visualization specialists have come together to create Stephen Hawking's 'Beyond the Horizon'.

Professor Hawking recently commented that he hopes that genetic engineering will be used to augment the human brain so that we can keep pace with the rapid advances in computers and remain superior to machines. Hawking is getting a voice upgrade - his new voice has been developed in India and will have an English Accent.

Larry King interviewed Hawking on his weekend program, and talked to him about a variety of subjects. He was called 'The Most Intelligent Man in the World'.

Professor Stephen Hawking uses modems to help him conduct research on the universe from the confines of his wheelchair. Hawking notes, "My body may be stuck in this chair, but with the Internet my mind can go to the ends of the universe."

MATHEMATICS

The most fundamental contribution of ancient India in mathematics is the invention of decimal system of enumeration, including the invention of zero. The decimal system uses nine digits (1 to 9) and the symbol zero (for nothing) to denote all natural numbers by assigning a place value to the digits. The Arabs carried this system to Africa and Europe.

The Vedas and Valmiki Ramayana used this system, though the exact dates of these works are not known. Mohanjodaro and Harappa excavations (which may be around 3000 B.C. old) also give specimens of writing in India. Aryans came 1000 years later, around 2000 B.C. Being very religious people, they were deeply interested in planetary positions to calculate auspicious times, and they developed astronomy and mathematics towards this end. They identified various Nakshatras (constellations) and named the months after them. They could count up to 10^{12}, while the Greeks could count up to 10^4 and Romans up to 10^8. Values of irrational numbers were also known to them to a high degree of approximation. Pythagoras Theorem can be also traced to the Aryan's Sulbasutras. These Sutras, estimated to be between 800 B.C. and 500 B.C., cover a large number of geometric principles. Jain religious works (dating from 500 B.C. to 100 B.C.) show they knew how to solve quadratic equations (though ancient Chinese and Babylonians also knew this prior to 2000 B.C.). Jains used Diameter. They were very fond of large numbers, and they classified numbers as enumerable, unenumerable and infinite. The Jains also worked out formulae for permutations and combinations though this knowledge may have existed in Vedic times. Sushruta Samhita (famous medicinal work, around 6th century B.C.) mentions that 63 combinations can be made out of 6 different rasas (tastes -bitter, sour, sweet, salty, astringent and hot).

MATHEMATICS

In the year 1881 A.D., at a village named Bakhshali near Peshawar, a farmer found a manuscript during excavation. About 70 leaves were found, and are now famous as the Bakhshali Manuscript. Western scholars estimate its date as about third or fourth century A.D. It is devoted mostly to arithmetic and algebra, with a few problems on geometry and mensuration.

With this historical background, we come to the famous Indian mathematicians.

Aryabhata (475 A.D. - 550 A.D.) is the first well known Indian mathematician. Born in Kerala, he completed his studies at the university of Nalanda. In the section Ganita (calculations) of his astronomical treatise, Aryabhatiya (499 A.D.), made the fundamental advance in finding the lengths of chords of circles, by using the half chord rather than the full chord method used by Greeks, that it was an approximation. (He gave it in the form that the approximate circumference of a circle of diameter 20000 is 62832.) He also gave methods for extracting square roots, summing arithmetic series, solving indeterminate equations of the type ax-by = c, and also gave what later came to be known as the table of Sines. He also wrote a text book for astronomical calculations, Aryabhatasiddhanta. Even today, this data is used in preparing Hindu calendars (Panchangs). In recognition to his contributions to astronomy and mathematics, India's first satellite was named Aryabhata.

Brahmagupta (598 A.D. - 665 A.D.) is renowned for the introduction of negative numbers and operations on zero into arithmetic. His main work was Brahmasputasiddhanta, which was a corrected version of old astronomical treatise Brahmasiddhanta. This work was later translated into Arabic as Sind Hind. He formulated the rule of three and proposed rules for the solution of quadratic and simultaneous equations. He gave the formula for the area of a cyclic quadrilateral. He was the first mathematician to treat algebra and arithmetic as two different branches of mathematics. He gave the solution of the indeterminate equation $Nx^2+1 = y^2$. He is also the founder of the branch of higher mathematics known as "Numerical Analysis".

After Brahmagupta, the mathematician of some consequence was Sridhara, who wrote Patiganita Sara, a book on algebra, in 750 A.D. Even Bhaskara refers to his works. After Sridhara, the most celebrated mathematician was Mahaviracharya or Mahavira. He wrote Ganita Sara Sangraha in 850 A.D., which is the first text book on arithmetic in present day form. He is the only Indian mathematician who has briefly referred to the ellipse (which he called Ayatvrit). The Greeks, by contrast, had studied conic sections in great detail.

Bhaskara (1114 A.D. - 1185 A.D.) or Bhaskaracharya is the most well known ancient Indian mathematician. He was born in 1114 A.D. at Bijjada Bida (Bijapur, Karnataka) in the Sahyadhri Hills. He was the first to declare that any number divided by zero is infinity and that the sum of any number and infinity is also infinity. He is famous for his book Siddhanta Siromani (1150 A.D.) It is divided into four sections - Leelavati (a book on arithmetic), Bijaganita (algebra), Goladhayaya (chapter on sphere -celestial globe), and Grahaganita (mathematics of the planets). Leelavati contains many interesting problems and was a very popular text book. Bhaskara introduced chakrawal, or the cyclic method, to solve algebraic equations. Six centuries later, European mathematicians like Galois, Euler and Lagrange rediscovered this method and called it "inverse cyclic". Bhaskara can also be called the founder of differential calculus. He gave an example of what is now called "differential coefficient" and the basic idea of what is now called "Rolle's theorem". Unfortunately, later Indian mathematicians did not take any notice of this. Five centuries later, Newton and Leibniz developed this subject. As an astronomer, Bhaskara is renowned for his concept of Tatkalikagati (instantaneous motion).

After this period, India was repeatedly raided by Muslims and other rulers and there was a lull in scientific research. Before Ramanujam, the only noteworthy mathematician was Sawai Jai Singh II, who founded the present city of Jaipur in 1727 A.D. This Hindu king was a great patron of mathematicians and astronomers. He is known for building observatories (Jantar Mantar) at Delhi, Jaipur, Ujjain, Varanasi and Mathura. Among the instruments he designed himself are Samrat Yantra, Ram Yantra and Jai Prakash.

SRINIVASA RAMANUJAM

This moving and astonishing biography tells the improbable story of India-born **Srinivasa Ramanujam**, self-taught mathematical prodigy. It all started when in 1913 Ramanujam, a 25-year-old clerk who had flunked out of two colleges, wrote a letter filled with startlingly original theorems in elegant English.

Srinivasa Ramanujam is undoubtedly the most celebrated Indian Mathematical genius. Srinivasa Ramanujam was born at Kumbakonam in a poor Tamil Brahmin family on December 22, 1887. His family then resided in the town of Kumbakonam, his father was a petty clerk in the local clothes shop. He attended school there and he was an inconspicuous, average student, till he asked his mathematics teacher about 'Zero' and 'infinity'. He questioned his teacher whether Zero divided by zero will mean 'one' like with any other number. His classmates laughed at his question, while his teacher was stunned at his genius. He explained to the other students the trick played by arithmetic, and that this simple-looking boy had asked a question that had taken mathematicians several centuries to answer. Some mathematicians claimed that zero divided by zero was zero. Others claimed that it was 'unity'. It was the Indian mathematician Bhaskara who proved that it was infinity. This boy, Srinivasa Ramanujam was a prodigy, and throughout his life-whether it was in Kumbakonam, or Cambridge, he was much ahead of not only his teachers, but also of any research, or findings.

While in school he came across a book entitled 'A synopsis of elementary results' in Pure and Applied Mathematics by George Carr. This book is just a compendium of results on integrals, infinite series and other mathematical entities found in analysis. Yet it left a lasting impression on Ramanujam; in fact it virtually determined his mathematical style. He would later write mathematics as a string of results without proof or with the barest outline of a proof. At the age of 13, Ramanujam was able to get Loney's Trigonometry from a college library. He not only mastered this book soon, but also started his own research with theorems, formulae not given in the book. This book triggered the mathematical genius in him. Largely self taught, he feasted on Loney's Trigonometry at the age of 13, and at the age of 15, his senior friends gave him Synopsis of Elementary Results in Pure and Applied Mathematics by George Carr. He used to write his ideas and results on loose sheets.

Senior students visited his sparse house to get their problems in mathematics solved. After school Ramanujam was hooked on mathematics. He spent all his time with his head over a slate working with problems in number theory that interested him and neglected everything else. The result was that he could never get through another examination. Mathematical ideas began to come in such flood to his mind that he was not able to write all of them down. He used to do problems on loose sheet of paper or on a slate and to jot the results down in the note books, as notebooks were a precious commodity those days. Before he went abroad he had filled three note books, which later became famous as Ramanujam's Frayed Note-books.

Although Ramanujam secured a first class in mathematics in the matriculation examination and was awarded the 'Subramanyan Scholarship', he failed twice in his first year arts examination and neglected other subjects such as history, English and physiology. This disappointed his father very much, and his parents decided to get him married forcibly to 'make him recover' from this madness of scribbling of numbers all the time. Eight-year-old Janaki became his wife, and now Ramanujam had to look out for a job to fend for himself and his wife. He had to find money not just to sustain them, but to also buy the papers needed to do his calculations. He needed about 2000 sheets of papers every month. To restrict the use of good papers, he started using even scraps of paper he found lying on the streets. Sometimes he used red pen to write over what was written in blue ink on the piece of paper he had picked up.

He visited the offices the whole day in search of a job. His appearance, over which he took scant care, and his frayed notebooks, to prove that he was good in accounts, put off people. Just as it happened to most geniuses, no one could understand his notebooks, and his application was turned down every where.

Luckily for him, at last he found someone who was impressed by his notebooks. The Director of Madras Port Trust, Francis Spring gave Ramanujam a clerical job on a monthly salary of Rs.25. Later some teachers interested in mathematics initiated a move to provide Ramanujam with a Fellowship.

On May 1, 1913, the University of Madras granted him a fellowship of Rs.75 a month, in spite of his not having the qualifying degree.

In the meantime Ramanujam kept showing his results to various people who he thought would be interested or would help him get a job that would give him a lot of time to do mathematics. He wrote to a couple of well known British mathematicians giving a list of some of the results he had obtained. They ignored him - thought he was a crank! Finally he wrote to one of the most distinguished English mathematicians of the time - a person who had done a lot of work on number theory - G.H. Hardy. He set out 120 theorems and formulae to Hardy, among them was what was known as Reimann series, a topic in the definite integral calculus. Though Ramanujam was ignorant of the work of the German mathematician, George F. Riemann, who had earlier arrived at the series as a rare achievement. Also included in the papers was a key formula in the hyper-geometric series, which later came to be named after him.

During the period between 1903 and 1914, Ramanujam worked in almost complete isolation in India. Throughout these years, he recorded his mathematical results without proofs in notebooks. Although many of his results were already in the literature, more were not.

In the meantime, Hardy, and his colleague J.E. Little Wood realised that they had discovered a rare mathematical genius. Hardy arranged for Ramanujam to come to Trinity College, Cambridge on March 17, 1914. He and Ramanujam met almost daily discussing mathematics for about three years.

With essentially no formal training, he managed to discover an enormous number of formulae, many of which were quite different from those previously known. The story of how he

SRINIVASA RAMANUJAM

came to the attention of the mathematical community and was brought into "formal" mathematical circles is an interesting one. Given the colonial nature of the relationship between England and India, it would have been "natural" for the English mathematicians to consider an untrained colonial native to be beneath them. However, to their credit, there is no evidence that the English mathematicians ever felt or acted this way.

Ramanujam found himself a stranger in Cambridge. The cold was too hard to bear, and being a Brahmin and a vegetarian, he had to cook his own food. However, his love for mathematics made him continue with his research with determination and fortitude. The constant company and appreciation of Hardy and Little Wood made it easy for him to bear much of the hardship he had to endure. Hardy found Ramanujam to be an unsystematic mathematician, similar to one who knows the Pythagoras theorem, but does not know what a congruent triangle means. Several discordance is found in his research, which could be attributed to his lack of formal education and training. Ramanujam played with numbers, as a child plays with a toy. It was his sheer genius that led him to mathematical "truths". The task of proving them, which is very important to science, fell on lesser mortals.

Ramanujam was elected Fellow of the Royal Society on February 28, 1918. He was the second Indian to receive this distinguished Fellowship. In October that year he became the first Indian to be elected Fellow of Trinity College, Cambridge. His achievements in Cambridge include the Hardy-Ramanujam-Little Wood circle method in Number Theory, Roger-Ramanujam's indentities in partition of integers, a long list of the highest composite numbers, besides work on the number theory and the algebra inequalities. In algebra his work on 'Continued Fractions' is considered to be equal in importance to that of great mathematicians like Leonard Euler and Jacobi.

While Ramanujam was immersed in his research, Tuberculosis, an incurable disease then was eating him up. He was sent back to India, exhausted, emaciated, and in agonising pain. He continued to play with numbers even in his death bed, and died shortly after at the age of 33 at Chetpet in Madras. He was also an astrologer, and a good orator, and he used to give lectures on "God, Zero and Infinity".

'The Man Who Knew Infinity: A Life of the Genius Ramanujam' was written by Robert Kanigel as a fitting tribute to this great soul.

Ramanujam had an intimate familiarity with numbers. During an illness in England, Hardy visited Ramanujam in the hospital. When Hardy remarked that he had taken taxi number 1729, a singularly unexceptional number, Ramanujam immediately responded that this number was actually quite remarkable: it is the smallest integer that can be represented in two ways by the sum of two cubes: $1729 = 1^3 + 12^3 = 9^3 + 10^3$.

The letters that Ramanujam wrote to G. H. Hardy on January 16 and February 27, 1913, are two of the most famous letters in the history of mathematics. These and other letters introduced Ramanujam and his remarkable theorems to the world and stimulated much research, especially in the 1920s and 1930s. This book brings together many letters to, from, and about Ramanujam. The letters came from the National Archives in Delhi, the Archives in the State of Tamil Nadu, and a variety of other sources. Helping to orient the reader is the extensive commentary, both mathematical and cultural, by Berndt and Rankin; in particular, they discuss in detail the history, up to the present day, of each mathematical result in the letters.

G.H. Hardy is at times in awe of Ramanujam's ability, proving to be a person of high quality as he tries as best he can to aid him in adapting to British society and to doing mathematics in a formal way. Many others are also involved, and it is clear that they do have a genuine interest in his welfare as a person and as a mathematician. There is absolutely no undercurrent of colonial class consciousness, given the situation in the early part of the twentieth century, that would have been typical of most English men of the times and it is a tribute to the mathematicians that they avoided it. Not many others would have been so noble.

Upon Ramanujam's death in 1920, G.H. Hardy strongly urged that Ramanujam's notebooks be published and edited. During the years 1903-1914, almost a decade after Ramanujam's death in 1920, the English mathematicians G.N. Watson and B.M. Wilson began this task in 1929, but although they devoted nearly ten years to the project, the work was never completed. In 1957, the Tata Institute of Fundamental Research in Bombay published a photostat edition of the notebooks, but no editing was undertaken. In 1977, Berndt began the tasks of editing Ramanujam's notebooks. Proofs are provided to theorems not yet proven in previous literature, and many results are so startling and different that there are no results akin to them in the literature.

This book is the second of four volumes devoted to the editing of Ramanujam's Notebooks. Part I, published in 1985, contains an account of Chapters 1-9 in the second notebook as well as a description of Ramanujam's second notebook. 'Ramanujam studies' have recently become something of an industry; several mathematicians are presently devoting their professional energy to the explication of inspired insight and unearthly beauty long buried in Ramanujam's notebooks. The present volume represents the continuation of a projected four-part series, by a mathematician associated with the University of Illinois. In six chapters it treats notebook entries pertaining to hypergeometric series, continued fractions, integrals and asymptotic expansions, infinite series, asymptotic expansions and modular forms. Chapters begin with an introductory essay, followed by notebook entries, proofs, corollaries and remarks, an overall objective being to detect and expose the pattern of Ramanujam's argument. If a result is known, the book provides references in the literature where proofs may be found; if a result is not known, the book attempts to prove it. Not only are the results fascinating, but, for the most part, Ramanujam's methods remain a mystery. Much work still needs to be done. Our hopes are with our young readers, who should strive to discover Ramanujam's thoughts and further develop his beautiful ideas.

President Dr. Abdul Kalam has suggested that instead of waiting for a Hardy to discover another Indian genius, we should set about seriously searching for one. It is a laudable suggestion from an eminent personality. It is not a question of creating a genius but of searching for one who may be lost in the oblivion of obscurity. India has witnessed a procession of spiritually realised souls, not in hundreds but may be in thousands over the millennia. Genius is the penultimate stage where, deviating from his own avowed goal of moksha, the Rishi diverts himself in a subject such as astronomy, mathematics, medicine, music, etc. The spiritual realisation in one generation, it is said, is never lost after that. It will be mostly in their descendents or those kindred souls that have picked up the vibration of Spirit. That blessed soul may not be aware of its presence in him. But it will never be totally suppressed and buried. It will show itself to the discerning eye, expressing itself in another unexpected field. All those who possess out of the ordinary talents in their own respective fields, whether it is tractor driving, building or public speaking, are potential candidates to be approached. The team that is on the lookout for the ONE must be given the training to look for extraordinary talents in a graded scale. On the surface it will be talent par excellence. Behind, it will lurk as essence of talent, a conceptual comprehension of the field hitherto unknown. Such people would be ridiculed in their own work spot. The potential candidates will be, more often than not, quiet and will be lost inside themselves. To frame all the criteria for such a search and reduce it to workable administrative procedures that are most likely to land on the target is possible. Should the parents suspect such a latent possibility in their child, the very best method to help it emerge is to teach the child utter truthfulness.

ASTRONAUT

KALPANA CHAWLA

KALPANA CHAWLA was the first and the only Indian-American woman to fly into space, Second Indian in space, after Indian citizen Rakesh Sharma who flew on a Soviet mission. One of a handful of Asian American astronauts and the only South Asian, Kalpana was married to Frenchman Jean-Pierre Harrison, a freelance flying instructor.

Born in Karnal, (Haryana) a small town in India, she graduated from the Tagore School in Karnal in 1976, and did a BS in Aeronautical Engineering from Punjab Engineering College in 1982. She moved to the US, picked up an MS in Aeronautical Engineering in 1984 from the University of Texas, and then a Ph.D in Aerospace Engineering from the University of Colorado, Boulder, in 1988. In 1988, Kalpana Chawla started working at NASA Ames Research Center in the area of powered-lift computational fluid dynamics. Her research concentrated on simulation of complex air flows encountered around aircraft such as the Harrier in "ground-effect."

When she attended the high school back in India, growing up, she thought that she was very lucky that she lived in a town which was very small but one of a handful of towns at that time which had flying clubs. She, along with her brother could see those small Pushpak airplanes, that students were flying as part of their training programs, which were not much different from Piper J3 Cubs that we see in the U.S. She and her brother, sometimes would be on bikes looking up, trying to see where these airplanes were headed. Every once in a while, they'd ask their dad if they could get a ride on one of these planes. And, he did take them to the flying club and get them a ride in the Pushpak and a glider that the flying club had. Kalpana thought that this was really her closest link to aerospace engineering that she could dig deep down her childhood and find out. Also growing up, she knew of this person, J. R. D. Tata in India, who had done some of the first mail flights in India, and also the airplane that he flew for the mail flights now hangs in one of the aerodromes out there that she had had a chance to see. For her, seeing this airplane and just knowing what this person had done during those years was very intriguing. Definitely it captivated her imagination. And, even when she was in high school if people asked her what she wanted to do, she told that she wanted to be an aerospace engineer. In hindsight, it was quite interesting to her that just some of those very simple things helped her make up her mind that, that is the area she wanted to pursue.

During her school years in India, she was forced by the educational system to figure out early what particular subjects she wanted to pursue. Since, she had to decide early on whether she was going to study science as in engineering or science as in medical, she decided to study physics, chemistry and mathematics as she was going to do aerospace engineering. And from then on, she was pretty much set on a track, and hoping. After pre-engineering, which is equivalent of 12th grade in US – she had started specializing in basic physics, chemistry, and mathematics and some languages, and was ready to go to an engineering college. She luckily got exactly what she wanted- a seat in aerospace engineering at Punjab Engineering College.

At that time her goal was to become an aerospace engineer. At that moment of time, becoming an astronaut was very farfetched for her. She wanted to pursue aircraft design. Though she had the wish to be a 'flight engineer', she did not have much idea about what exactly a flight engineer did. It was sheer coincidence that later she ended up doing what she studied about early on the subject.

She thinks that inspiration is tied up with motivation, and for her, it came from every day people from all walks of life. It was easy for her to be motivated and inspired by seeing somebody who just goes all out to do something. For example, some of her teachers in her high school, the amount of effort they put in to carry out their courses, the extra time they took to do experiments with the students, all that never failed to motivate her. She was thrilled by the compliments the teachers gave to the students to come up with ideas - new ideas, and she often wondered how they even had the patience to look at these simple things and spare a word or deed of appreciation.

She held a Certificated Flight Instructor's license with airplane and glider ratings, Commercial Pilot's licenses for single- and multi-engine land and seaplanes, and Gliders, and instrument rating for airplanes. She enjoyed flying aerobatics and tail-wheel airplanes.

In general in her life, explorers had inspired her. She was fond of reading about them, about Shackleton, the four or five books written by people in more recent times, and then during the expedition. And about some of the incredible feats these people carried out; like making it to the Pole almost, but making the wise decision to stop a hundred miles short and return. Lewis and Clark's incredible journey across America to find a route to water, if one existed. And, the perseverance and incredible courage with which they carried it out. About Patty Wagstaff, on how she started flying aerobatic airplanes, who later on became an unlimited U.S. champion three times in a row.

She was also inspired by so many people out there who had done some incredible things., In explorers, she was inspired by

Peter Matthiessen and how he has explored the whole world and chronicled life, of both animals and birds as they exist. And, also because he did it by simply walking everywhere on his feet. People who went and climbed the Himalayas, travelled across Africa. When she read about these people, the one thing that just stood out was their perseverance in how they carried out what they wished to carry out.

Her research concentrated on simulation of complex air flows encountered around aircraft such as the Harrier in "ground-effect." Following completion of this project she supported research in mapping of flow solvers to parallel computers, and testing of these solvers by carrying out powered lift computations. In 1993 Kalpana Chawla joined Overset Methods Inc., Los Altos, California, as Vice President and Research Scientist to form a team with other researchers specializing in simulation of moving multiple body problems. She was responsible for development and implementation of efficient techniques to perform aerodynamic optimization

Selected by NASA in December 1994, Kalpana Chawla reported to the Johnson Space Center in March 1995 as an astronaut candidate in the 15th Group of Astronauts. After completing a year of training and evaluation, she was assigned as crew representative to work technical issues for the Astronaut Office EVA/Robotics and Computer Branches. Her assignments included work on development of Robotic Situational Awareness Displays and testing space shuttle control software in the Shuttle Avionics Integration Laboratory.

In November, 1996, Kalpana Chawla was assigned as mission specialist and prime robotic arm operator on STS-87 (November 19 to December 5, 1997). STS-87 was the fourth U.S Microgravity Payload flight and focused on experiments designed to study how the weightless environment of space affects various physical processes, and on observations of the Sun's outer atmospheric layers. Two members of the crew performed an EVA (spacewalk) which featured the manual capture of a Spartan satellite, in addition to testing EVA tools and procedures for future Space Station assembly. In completing her first mission, Kalpana Chawla travelled 6.5 million miles in 252 orbits of the Earth and logged 376 hours and 34 minutes in space. In January, 1998, Kalpana Chawla was assigned as crew representative for shuttle and station flight crew equipment. Subsequently, she was assigned as the lead for Astronaut Office's Crew Systems and Habitability section.

Her love of flying, and her exposure to a wide variety of computer systems at the NASA Ames Research Center in California helped her in her mission. She became a naturalized US citizen, and got married to Jean-Pierre Harrison, a freelance flying instructor.

Chawla's space flight in 1997, the first by an Indian-American, made her a powerful symbol of achievement. She provided inspiration throughout India, where her missions were front page news. The Indian diaspora, revered her as a role model.

The 16-day flight on STS-107 Columbia January 16 to February 1, 2003 was a dedicated science and research mission. Working 24 hours a day, in two alternating shifts, the crew successfully conducted approximately 80 experiments. The STS-107 mission ended abruptly on February 1, 2003 when Space Shuttle Columbia exploded over Texas, at a height of 2,00,000 feet, 16 minutes prior to scheduled landing.

Lost in that explosion were seven lives, including that of Kalpana Chawla, 42, a horrific end to a life that had had its genesis in horror of a quite different kind. Kalpana's story is incomplete without the story of her parents, especially that of Ban Lal Chawla, who landed in the wilderness of Karnal a few days after August 15.

We should go a little backwards into history and pay attention to the life of Banarasi Lal Chawla, Kalpana's father to know where she got her 'do or die' spirit from.

Chawla's exodus from the dusty outskirts of Lahore into northern India where he rebuilt his life was to peak when his daughter Kalpana became an astronaut. Chawla saw Kalpana's achievement as vindication, as the final sign that the wounds of Partition had healed.

A few days earlier, as news of the massacres of Hindus in Pakistani villages began pouring in, Chawla, his mother, two brothers, and a sister had moved to Choorkana Mandi from their village Shehupura. With his father away in Bikaner on work, it was left to Chawla to lead the family's exodus from their ancestral village.

It was sultry, and dark, on an August evening in 1947 when Banarsi Lal Chawla, then 14, lay on a railway track, thirsty, hungry, unconscious, and bleeding. Around him, open coal wagons echoed with the cries of children, most of whom were living the final hours of their lives. Mr. Chawla remembers that day in August 1947, when a fifth of the world's population was convulsed by Partition and forced to flee their homes. How people were packed into the open wagons like potatoes. Hardly had it moved a few kilometers, the train was stopped at Shahdra, on the outskirts of Lahore, while his whole family, including his mother, were sitting inside the wagon, he had to be content with a perch on top of a joint between two wagons. That is where he got a place.

As the warm morning made way for a blistering hot noon and faded to evening, people began a desperate search for water.

Food, by then, was a luxury not even thought of. Mr. Chawla joined hundreds of other men, women and children lining up to sip the dirty water that had filled the pits near the track- water from the rains the day before.

He then returned to his perch, and that is all he recalls. Around 10 pm, his uncle found Chawla unconscious, precariously close to the wagon wheel — he had fallen off his perch in his stupor. His uncle took him to yet another pit, gave him more dirty water to drink, and washed the deep gash in the same water. Chawla returned to his perch, his feet dangling down the side of the wagon's cabled joint, and continued his vigil along with several hundred others. As night progressed, a mob that had gathered began firing, with the intent of avenging itself for the killing of Muslims in India. One bullet whistled past Chawla, brushing his ears. That hiss of death passing within inches, remains a landmark sound among the many that comprise the noisy, eventful, maverick life that Chawla went on to lead.

Despite moving to the safety of his uncle's house in Choorkana Mandi, Chawla couldn't get over the thought of his cattle, which were left unattended in the village. So he coerced his uncle to accompany him to their village to rescue the cattle. En route, an acquaintance met them and warned of mass killings, and pleaded with them to go back. Chawla's uncle sent the boy back and went forward on his own. He never returned.

Chawla's father had been awaiting his family for days at the Amritsar railway station. It was a hopeless wait, since a group of refugees from Lahore had told him that his children and brother had been killed. Later, he was told they were alive — and he did not know what to believe. On August 18, at about 2 am, when the open coal wagons entered into Amritsar carrying hundreds of refugees, many of them dead, Chawla was into his sixth day of waiting. The family, now reunited, took a train leaving for Delhi, then the ultimate destination for the millions of refugees fleeing Pakistan. The Chawlas, the extended family at this point numbered 20 preferred to hop off at Karnal in Haryana, some 130 kilometres from Old Delhi.

The family moved into the first available vacant building: a mosque approximately 15x18 feet, with no doors and just a dirty well in one corner. Chawla and his father set out to seek food. Chawla recalls cleaning up the 60-foot well and searching for a job, while his father teamed up with a relative and set up a small shop. However, the strain told on the elder Chawla, who fell ill a few months after the family settled down in Karnal.

For Chawla junior, that was when life began in earnest. When he talks of those days, he is emotionless. It is almost as if, having seen it all, having endured it all, he can no longer be roused by mere memories.

It was much later that he finally found his niche – Chawla went back to making boxes, this time for the hordes of refugees who had thronged Karnal with nothing to store the rations the government was distributing. He began selling five to 10 boxes a day, and the business boomed as shops too began demanding his wares. Shortly thereafter, he married Sanjogta Kharbanda, the educated daughter of a doctor who, too, had fled the horrors of Partition. As the business prospered, the family expanded — daughters Sunita and Deepa came first, then son Sanjay, then the baby of the family, Kalpana, in 1961.

Next came the Binny Showroom, which was a major success, and then came a company for 'tyres'. Meanwhile, his children were growing up, and proving to be intelligent. In fact, Chawla saw nothing special about Kalpana, in that respect, eldest daughter Sunita is a gold medalist from Punjab University.

Son Sanjay joined the Karnal flying school, and Kalpana, engineering college. Ironically, by then Chawla was so busy, her father was unaware that his youngest daughter had opted for aeronautical engineering, of absolutely no use to the owner of a flourishing tyre business. He expected her to join his tyre business.

During a break from studies, Kalpana accompanied her brother to flight school, but the authorities demanded her to get the written consent of her guardian. Chawla refused consent. As Sunita remembers it, Sanjay was to give Kalpana some valuable advice: "Everyone fights his own battles."

Chawla was in the US when Kalpana learnt that she had topped Punjab University in the engineering finals, and was offered a job in her own college. But she had already begun applying to several American universities, and was accepted by the University of Texas for a Master's in aeronautical engineering.

Her father was away and in the male-dominated household, no one else could take a decision. So Kalpana went back to Punjab Engineering College and took up a teaching job. "I returned after two months and reached Karnal late one evening," Chawla recalls. "Kalpana was supposed to be home, but she wasn't. I asked about her. She is in Chandigarh, I was told. And then, someone said, anyway why are you asking? You don't have time for her." It triggered a family revolt, with his wife, whom Chawla calls "liberal and advanced" and the three elder children ganging up on behalf of their baby of the household. "I asked them what she was doing in Chandigarh. They said, why don't you go and find out?"

Early next morning, on August 26, Chawla reached Kalpana's hostel in Chandigarh, but she wasn't there. So he went to the college to visit the principal, whom he knew. "Chawla, you have only money, nothing else," the principal said, and told the astonished father about how brilliant Kalpana was, and that time was running out if she was to get into a US college. Chawla and the principal walked over to where Kalpana was taking classes. "She was writing on the blackboard, with her back to the class. After a while, she turned, wiping the chalk dust off her hands, and as she turned, she saw me.

"She walked up to me in tears and said, Papa, you have destroyed my career. You never have time for me."

The date was August 26 — and the last date for admission to Texas was the 31st of that month. Kalpana had no passport, no visa, no tickets, nothing.

Chawla cried, tears of genuine distress. And through his tears he asked his daughter, "Do you want to go to the US?"

"Yes. I will go on my own money," Kalpana replied.

"You can do that, but I can fund you, as well," Chawla said.

"Anyway, now I can only go next year," his dejected daughter said. "I have no passport, no visa, nothing."

If his life had taught him one thing, it was to never give up. "Do you want to go this year?", Chawla asked his daughter.

"Yes", Kalpana replied.

"Then come with me," he said, asking her to resign from her job that instant.

Kalpana was reluctant, fearing that her father would force her to join Super Tyres. "She thought I was trying to trick her into coming back to Karnal, and once there, I wouldn't let her leave," Chawla recalls.

Pulling every string he knew, drawing on all his accumulated goodwill, Chawla got his daughter's passport the same day. A day later, the visa was organised. On August 28, Kalpana, accompanied by brother Sanjay, boarded a British Airways flight at midnight.

The story was to take another twist, when the flight was first delayed, then cancelled. The Chawla family, which had gone to see Kalpana off, was in tears. But Chawla, even then, did not know the meaning of failure. He began calling friends in the US, and finally arranged for Kalpana to be admitted behind schedule. In fact, the university even organised a pickup for Kalpana and her brother from the airport.

Shortly before Kalpana took off on her ill-fated last flight her father, now deeply into religion, philosophized about his youngest daughter's achievements. "Good things happen in families where good people are born," he said. He recalled how, when his father Lala Labhamal was around 45 years old and still struggling to establish himself after the trauma of Partition, he met a guru and became his disciple. He built a math in Karnal and ran it till he was 85. He died in 1997 — the same year that Kalpana took off on her maiden space sojourn. 'Nirmal Kutiya' continues to be run by his disciples, providing succour to Karnal's poor.

The family tradition of serving society is now being carried on by Chawla's younger brother, Amrit Chand Chawla. The industrialist from Mumbai has left his factory to managers and spends his time in Karnal, where he runs a well-furnished old age home for some 160 people, and a school where around 2,000 poor children are provided education and basic necessities free. He also provides some 700 poor families a monthly allowance to meet their needs.

Meanwhile, a second generation was growing up, and taking inspiration from Kalpana's odyssey from Karnal to outer space. Megha, a standard five student, told this correspondent shortly before Kalpana took off on what was to be her last voyage, that she wanted to be an astronaut like her aunt.

Till the evening of Saturday, February 1, the story was pure Horatio Alger, a man who survived untold horrors and went on to make a fortune; and his daughter who, against the odds, went on to make her name in one of the most challenging of careers.

Today, that daughter's life, her achievements, ended in a fireball that destroyed her spacecraft. And left behind, by that explosion, is an old man who, finally, finds a tragedy too great for even his innate stoicism to withstand.

On Friday, Satveer Chaudhary mailed a letter addressed to Kalpana Chawla to congratulate her on her second trip to space and to tell her how much she means to Indian-Americans, including those in the Twin Cities. The letter never reached her. Chawla and six other astronauts died on Saturday when the space shuttle Columbia disintegrated over Texas.

It is just an unbelievable career that Kalpana had, if someone can immigrate to the United States in the 1980s and 10 years later represent the U.S. in space, she proved that anyone can reach the American Dream and literally reach for the stars.

She was kind of an icon for the immigrants from India. She proved that a girl from Karnal, a small town, could make it to the top. Chawla's hometown of Karnal remembers with pride her first spaceflight in 1997. Chawla was this little girl who once sat in her school, and went on to become an Indian pioneer. But her ascent from small-town India to the weightlessness of space was met with some resistance from a culture in which women traditionally do not have careers outside of marriage and motherhood. But Chawla's relentless demands and her mother's eventual support finally earned her freedom.

On Sunday, she was mourned like a hero in Karnal.

Hundreds of people stopped by her childhood home and gathered at her high school and elsewhere. They prayed at makeshift shrines where incense burned in front of her photograph draped in garlands of marigolds. Joy Sidhu, a schoolteacher, brought her 15 year-old daughter and other girls to one of the memorials.

They stood at the edge of the crowd, watching the crush of local officials, nearly all men, push to have their pictures taken while they laid wreaths.

"I brought them here so I could say 'Look, there are options for you,' "Sidhu said of her students. "There is marriage and there are children, but there are other choices as well."

As the world mourned, Indian-Americans in the Twin Cities honoured Chawla at a memorial service at 7 p.m. Mournful Tuesday at a temple in Indianopolis, USA.

AGRICULTURE

Nature's essential wisdom is shown in its character and its bounty, existed much before the advent of human intellect, through which life evolved with all its splendour, variety and balance. To enhance an ancient civilization like ours, countering the burgeoning population on one side, and the depleting, degrading environment on the other, we need much more than all these. We need wisdom and genius to combine them with scientific knowledge, which in turn guides the technological forces we have turned loose in the past century. The answer has come from people like Dr. M.S. Swaminathan, teacher, researcher, administrator, whose range of achievement is spectacular, inspiring and a resounding legacy for generations to come. He crusades endlessly for scientific improvements in agriculture, analyses ways for cheaper and bountiful production of food crops, and last but not the least, its modernization.

M.S. SWAMINATHAN

Monkombu Sambasivan Swaminathan, the 'Sun of the Soil', was born on August 7, 1925 at Kumbakonam and had his basic education in Tamil Nadu. He came from a family of doctors and freedom fighters. His father Dr. M.K. Sambasivan was a surgeon and as the Municipal Chairman of Kumbakonam, eradicated a particular form of filarial mosquitoes in as early as the thirties. He takes great pride in belonging to the particular Gandhi-Nehru generation, imbibing the spirit of nationalism, obligation and loyalty to the country, but he was moved by the agony of watching our country rocked by the after-effects of second World War, and the terrible Bengal famine. People died like flies for want of food. The dying millions had a lasting effect on his conscience, and he resolved to work towards agricultural (food) prosperity of his country. His aim saw him through a B.Sc. in agriculture from the college in Coimbatore. Then further on to studying Genetics and Breeding in the Agricultural Research Institute in New Delhi, a UNESCO Fellowship to Holland for further studies. He then went to Britain and took his Ph.D. from the School of Agriculture in Cambridge in 1952. He worked for a year in the university of Wisconsin as Research Associate. Throughout his life the only aim was to equip himself, and return to serve his country. In 1954, he returned to a research post in Cuttack Central Research Institute, where he worked on breeding fertilizer-responsive varieties of rice. He spent the next two decades doing research on various crops and in basic applied genetics at the Indian Agricultural Research Institute.

This was the time, the late fifties, when India was facing a challenge in food self-sufficiency. With the traditional methods of farming, the future prospects of feeding its teeming millions were becoming very bleak. Always this young and brilliant scientist felt that he has to rise up to meet this need of his mother country. After persistent research he felt that Noble Laureate N.E.Borlaug's newly developed Mexican dwarf wheat variety could solve the wheat problem of our country. On Swaminathan's initiative, Borlaug visited India, and he suggested a range of the Mexican Dwarf variety suitable for our climatic and cultivating conditions. Wheat production increased dramatically, and within a decade production had doubled. He also developed high yielding strains of wheat and rice and crosses in potato and jute species. Thus, Dr. Swaminathan was responsible for the Green Revolution in our country.

He feels that we are adequate on the food front, and poverty is the cause of hunger, not insufficient food reserves. The governments have more than 30 million food reserves on hand, yet our people die of hunger, and ours is an induced hunger. Our farmers are poor, each holding half to one acre of land compared to about 200 acres in the developed countries. We have landless labourers whose take-home pay is a meagre few hundreds of rupees a month. Alcoholism is another evil that stalks the farming community. According to him, financial access, education, enlightenment, and economic empowerment are more important than cultivating more. Our farmers can do much better if they are given more buyers and a better price. He feels that dry land farming and drought management needs much more investment than the normal one. He advocates the formation of 'Southern Water grid' on the lines of 'Power grid', linking the major Southern rivers, thereby pooling the water that is squandered into the sea. He hates the inaction by the government on this score, and calls the schemes only on paper as 'paralysis by analysis'. He is working towards 'biological control' with minimum essential chemical pesticides and integrated pest management with plant pesticides with garlic and neem spray combined with pyrethrum and tobacco decoction. He believes that water resources and the land use hold the key to any future prosperity in the field of agriculture. According to him, fertile lands being diverted to house-building, brick kilns, urban invasion in the form of real estate poses a grave danger to farming. Compared to other well-developed Asian countries like Japan and China, there is a very big gap between our capacity and final output. So, he advocates for a very comprehensive, compatible, circumspect land use for India's future prosperity. He also feels that there is a real danger from deforestation, which the country should tackle on an emergency basis. The 35% of the forest areas in 1947 has been reduced to a scanty 12 % by 2000, the rest being degraded forest! We should concentrate to eradicate the population pressure on the natural resources, the inability of the soil to regenerate, the affected hydrological cycle of the water, and arrive at a consistent water strategy. He feels that more than the government, the drive, the motivation must come from the people, as 'Greening India', 'Participatory forest management', 'Preservation of forest', are all 'people's movement', like the social fencing in china. Only movements that involve people's participation in a sustained and regular manner, is bound to have long term results. He also has a very fine estimation of our agricultural colleges. He feels they are the best in the country. Yet, what is lacking is the application. He advises the educated youth not to go in search of 'white collar jobs', but take to 'knowledge intensive ecological farming'. He believes in Gandhiji's message- 'Indian agriculture will continue to stagnate unless we reverse the drain of brain from the villages to the cities'. Application and dedication to improve the quality of farming is what is needed for now.

As regards the highly contentious issue of 'patent rights', he agrees that to protect our rights, we need empowerment through legislation. Any agricultural policy of the government should be pro-farmer, pro-poor, and pro-women for it to be effective in our country. The IT industry should be used to advance the standard of living of our farmers, which in turn would lead them to compete with the 'best in the west'!

Added to being a great scientist, he is also an able administrator and a person who is keenly interested in uplifting the living conditions of his people. He is especially interested in uplifting the working conditions of the rural folk especially women. At the forefront of his research is for the modernizing agriculture, thereby increasing productivity. He formulates various projects and gets the research done in his laboratories, the benefit of which, he makes sure, must reach the farmer and get implemented in his field.

M.S. SWAMINATHAN

In 1971 he was awarded the Ramon Magsaysay Award for generating the hope that our country can become self-sufficient in food products. In 1973 he was elected Fellow of the Royal Society. He has also received the S.S. Bhatnagar Award, the Birbal Sahni Medal and the Mendel Memorial Award. He held the post of Director, International Rice Research Institute, Philippines. Swaminathan was also the first agriculture scientist to win the Albert Einstein World Science Award in 1986. At present he is the Director of the M.S. Swaminathan Research Foundation, Chennai.

Prof. M.S. Swaminathan has been acclaimed by TIME magazine as one of the twenty most influential Asians of the 20th century and one of the three from India, the other two being Mahatma Gandhi and Rabindranath Tagore. He has been described by the United Nations Environment Programme as "the Father of Economic Ecology" and by Javier Perez de Cuellar, the former Secretary General of the United Nations, as "a living legend who will go into the annals of history as a world scientist of rare distinction". He was Chairman of the UN Science Advisory Committee set up in 1980 to take follow-up action on the Vienna Plan of Action. He has also served as Independent Chairman of the FAO Council and President of the International Union for the Conservation of Nature and Natural Resources.

A plant geneticist by training, Prof. Swaminathan's contributions to the agricultural renaissance of India have led to his being widely referred to as the scientific leader of the green revolution movement. His advocacy of sustainable agriculture leading to an ever-green revolution makes him an acknowledged world leader in the field of sustainable food security. The International Association of Women and Development conferred on him the first international award for significant contributions to promoting the knowledge, skill, and technological empowerment of women in agriculture and for his pioneering role in mainstreaming gender considerations in agriculture and rural development. Prof. Swaminathan was awarded the Ramon Magsaysay Award for Community Leadership in 1971, the Albert Einstein World Science Award in 1986, the first World Food Prize in 1987, Volvo Environment Prize in 1999, and the Franklin D Roosevelt Four Freedoms Award in 2000.

Prof. Swaminathan is a Fellow of many of the leading scientific academies of India and the world, including the Royal Society of London and the US National Academy of Sciences. He has received 43 honorary doctorate degrees from universities around the world. Recently, he has been elected as the President of Pugwash Conferences on Science and World Affairs. He currently holds the UNESCO Chair in Ecotechnology at the M.S. Swaminathan Research Foundation in Chennai (Madras), India.

In 2004, he has been asked to head the National Commission on farmers', to assess the condition of Indian agriculture, accelerate the reform process with special reference to quality production and marketing, identify the factors responsible for the imbalances and disparities in their living conditions, suggest measures to improve them, and review the different conditions of farmers, in various regions. He will prepare a National policy on Agri-bio technology. India could not have found a better man to shoulder the onerous responsibility.

THE NOBEL LAUREATES

SIR C.V. RAMAN

Chandrasekhar Venkata Raman, popularly known as C.V. Raman, was born in Thiruchirapalli a town, on the banks of the river Cauvery, in Tamil Nadu, India on November 7, 1888. He was the second child of Chandrasekhara Iyer and Parvathi Ammal. Chandrasekhara Ayyar was a scholar in Physics and Mathematics. He loved music. His wife was Parvathi Ammal. Their second son, a child genius was named as Venkata Raman. He was also called Chandrasekhara Venkata Raman or C.V. Raman. His father was a  professor of Mathematics. At an early age, Raman moved to the city of Visakhapatnam, in the present day state of Andhra Pradesh, where his father accepted a position at the A.V.N. College. Raman grew up in an atmosphere of music, Sanskrit literature and Science. Raman's academic brilliance was established at a very young age. At eleven, he finished his secondary school education and entered A.V.N. College and two years later moved to the prestigious Presidency College in Madras (present name, Chennai). When he was fifteen, he became Bachelor of Arts (B.A.) with honours in Physics and English. He stood first in every class and was talked about as a child genius. He joined the M.A. class in the same college and chose Physics (study of matter and energy) as the main subject of study. Love of science, enthusiasm for work and the curiosity to learn new things were natural to Raman. Nature had also given him the power of concentration and intelligence. He used to read more than what was taught in the class. When doubts arose he would set down questions like 'How?' 'Why?' and 'Is this true?' in the margin in the textbooks.

During that time students who did well academically were typically sent abroad (England) for further studies. Because of Raman's poor health he was not allowed to go abroad and he continued his studies at the Presidency College. In 1907, barely seventeen, Raman again graduated at the top of his class and received his M.A. with honours. In the same year he married Lokasundari. He adored science and at 19 he became a member of the Indian Association for Cultivation of Science, the headquarters of our country's pioneering scientific body.

Professor Eliot of Presidency College, Madras, saw a little boy in his B.A. Class. Thinking that he might have strayed into the room, the Professor asked, "Are you a student of the B.A. class?" "Yes Sir," the boy answered. Your name?, the professor asked. "C.V. Raman." he replied. This little incident made the fourteen-year-old boy well known in the college. The youngster was later to become a world famous scientist.

At the time of Raman's graduation, there were few opportunities for scientists in India. This forced Raman to accept a position with the Indian Civil Services as an Assistant Accountant General in Calcutta. There, he was able to sustain his interest in science by working, in his spare time, in the laboratories of the Indian Association for the Cultivation of Science. He studied the physics of stringed instruments and Indian drums.

The works of the German scientist Helmhotlz (1821 - 1891) and the English scientist Lord Raleigh (1842 - 1919) on acoustics (the study of sound) influenced Raman. He took immense interest in the study of sound. When he was eighteen years of age, one of his research papers was published in the 'Philosophical Magazine' of England. Later another paper was published in the

SIR C.V. RAMAN

scientific journal 'Nature'. He began to write research papers for reputed science journals, and though he took up an administrative job in the Finance Ministry in Calcutta, his interest in science was at its peak. He spent hours after office in laboratory of the association, working throughout the night.

Sir C.V. Raman was interested in acoustics- 'the science of sound'. He studied how the bowed string instruments like violin and the sitar could produce melodious music. He gave lectures on the acoustics of the violin in London, when a scientist asked him if he was interested in becoming a Fellow of the Royal Society by dabbling in Physics, a subject that was close to his heart. That set him thinking, and the passion that was lying deep in his heart for science and the unanswered questions of the mysteries of the world fascinated him. His first trip outside India was to Oxford in 1921 to represent the University of Calcutta. On his return journey from London he sought answers for the many wonders of the world, including the blueness of the sky. He concluded that the sky was blue because of the scattering of light by water molecules. He felt his intuition was right, and hastened to his laboratory in Calcutta to prove the theory right. Thus began the research in optics that was to make him famous as an universally known scientist.

In 1924 Raman was elected Fellow of the Royal Society for his contributions to optics. In 1927, when he was deep into his research on his theory of Light in his laboratory, that had no sophisticated equipment, he heard that Professor Crompton has won the Nobel Prize for his theory on the nature of X-Rays when passed through matter. The change was dependent on the kind of matter. This effect was called the 'Crompton effect', and this excited Raman as he felt that if 'Crompton Effect' was true, so was his theory on light. He felt that if light also changed its nature when passed through a transparent medium, then his five years of research on the optics, the science of light was correct. He felt that he could find the answer with some modifications in his equipment.

During his voyage, he conducted some experiments and published a note in Nature entitled 'The Colour of the Sea'. It was a generally held belief that the blue colour of the sea is due to the reflected sky-light as well as due to absorption of the light by the suspended matter in the water. Raman showed that the blue colour of sea is independent of sky reflection as well as absorption, but rather it is due to the molecular diffraction. These initial experiments opened up a new field of research in Calcutta. Further work on the scattering of light led to the discovery of the Raman effect in 1928. The effect deals with the change in the frequency of the monochromatic light after scattering. The spectrum of the scattered light gives clues about the molecular structure of the material under study, thereby helping to understand its properties.

On March 16th, 1928, Raman announced his discovery of 'new radiation' to a convention of scientists in Bangalore. The world hailed the breakthrough as "Raman Effect". With equipment worth Rs.200, and an ill-equipped lab, with not much of previous research to guide him, Raman magnetized the world's attention on our country. 'Raman Effect' brought the 1930 Nobel prize in Physics to C.V. Raman.

The 'Raman Effect' is the phenomenon that causes changes in the nature of light when it is passed through a transparent medium, whether solid, liquid or gaseous. The phenomenon takes place when molecules of the medium scatter light energy particles, "photons", just as a striker scatters a bunch of coins on a carrom board. From the minute changes observed in the energy of photons, or nature of light, the internal molecular structure of the medium can be deduced.

SIR C.V. RAMAN

The "Raman Effect" is important in understanding the molecular structure of chemical compounds. In fact, within a decade of its discovery, the internal structures of some 2,000 chemical compounds were determined. Subsequently, the internal structures of crystals were also determined. With the invention of the "laser, with its powerful light radiation, the "Raman Effect" has become a powerful tool for the scientists.

He advised young scientists to have an inquisitive mind and watch the world around them, rather than confining themselves to their laboratories. "The essence of science is independent thinking and hard work, not equipment," he said.

In 1917, with his scientific standing established in India, Raman was offered the position of Sir Taraknath Palit Professorship of Physics at Calcutta university, where he stayed for the next fifteen years. During his tenure there, he received world wide recognition for his work in optics and scattering of light. He was elected to the Royal Society of London in 1924 and the British made him a knight of the British Empire in 1929. The following year he was honoured with the prestigious Hughes medal from the Royal Society. In 1930, for the first time in its history, an Indian scholar, educated entirely in India has received the highest honour in science, the Nobel Prize in Physics.

Raman's work on musical instruments was well known outside India even before he joined the University of Calcutta as a professor. He began his work on light scattering in 1921 and soon established his reputation in this field. In 1924 he was invited by the British Association for the Advancement of Science in Toronto to open a discussion meeting on the scattering of light. As a scientist, he established several contacts with the scientists in the West.

In 1930, C. V. Raman was the first 'non-white', Asian and Indian to receive the Nobel prize in physics for his work on scattering of light and discovery of the Raman effect. The documents were obtained from the Nobel Committee connected with the proposal and selection of C. V. Raman for the Nobel Prize and the results of the studies are reported in this paper.

In 1934, Raman became the director of the newly established Indian Institute of Sciences in Bangalore, where two years later he continued as a professor of physics. He was interested in acoustics and discovered that 'mridangam' and tabla unlike other drums, possess harmonious overtones. He was fascinated by anything that was colourful, be it a butterfly, gem or a flower. His mind was forever searching answers to the enigmatic wonders of the world. Why do some things look beautiful? What makes gems and stones bright and colourful? He was forever trying to find answers.

In 1947, he was appointed the first National Professor by the new government of Independent India. He retired from the Indian Institute in 1948 and a year later he established the Raman Research Institute in Bangalore, served as its director and remained active there until his death on November 21, 1970, at the age of eighty two. Raman was honoured with the highest award, the "Bharat Ratna" (Jewel of India), by the Government of India.

The genius with simple equipment barely worth Rs. 300, was the first Asian scientist to win the Nobel Prize, and he won it for his astounding research findings in Physics. He was a man of boundless curiosity and a lively sense of humour. His spirit of inquiry and devotion to science laid the foundations for scientific research in India. And he won honour as a scientist and affection as a teacher and a man.

A short life sketch of the founder and the foundation of the Nobel Prize is included in Sir C.V. Raman's biography to help the reader to realise the magnitude and exclusivity of this prize.

The Nobel Prize is one of the prizes known to a great part of the non-scientific public to be the highest honour to be awarded to scientists. The Statutes (or the ruling) of the Nobel Foundation (SNF) which were approved by the Crown on 29 June 1900 had been decreed by the Swedish Government on 27 April 1995. The rules and regulations quoted here are taken from these statutes.

The list of honours bestowed on Raman for his scientific findings is long, but the best was undoubtedly the 'Nobel Prize'. Raman received the Nobel prize in a record time of two years after his prize- winning discovery. Several questions have been raised about not sharing of the prize by Raman either with his colleagues or the Russian scientists. It will be shown here that it was not in Raman's hand to take this decision. The reasons for these are elaborated here.

However discussing of the reasons for his being awarded the Nobel Prize, as well as some of the questions that have been raised on his receiving of this award, a short biography of the founder of the Nobel prize has also been included. This brief is included in Sir C.V. Raman's biography as a proof of his genius, awesome pioneering work and his reputation among the scientists of these countries.

The relevance of this discovery in the area of quantum mechanics can be judged from the statement of R. W. Wood, 'It appears to me that this very beautiful discovery which resulted from Raman's long and patient study of the phenomenon of light scattering is one of the best convincing proofs of the quantum theory'.

Alfred Nobel was born on 21 October 1833 in Stockholm, Sweden. His father Emmanuel Nobel was an engineer who built bridges and buildings. In connection with this, he experimented with different techniques for blasting rocks. In 1837, he had to declare himself bankrupt. He left Stockholm and moved to Russia. In 1842, the rest of family joined him in St. Petersburg, where Alfred and his brothers were tutored privately till 1850. After the Crimean War (1853-1856), once again, Emmanuel Nobel had to declare himself bankrupt and he returned to Sweden in 1859.

In Paris, Alfred Nobel worked in the private laboratory of a famous chemist T. J. Pelouze, and came in contact with an Italian scientist, A. Sobrero, an inventor of highly explosive nitroglycerine. This idea was extended further by Alfred Nobel to conduct explosions under controlled conditions. After a long period of experimentation he was able to turn liquid nitroglycerine into a ductile explosive and patented this material as dynamite in the year 1867. He also invented a detonator which could be ignited with a fuse. These inventions helped reduce the costs for drilling tunnels, building canals and other construction works.

At the end of his life, he had as many as 355 patents. Some of his industrial enterprises still exist, e.g. Imperial Chemical Industries, UK; Dyno Industries, Norway; and AB Bofors, Sweden. Through his skill as industrialist, and his number of patents he became one of the wealthiest men in the world. Alfred Bernhard Nobel died in Italy on 10 December 1896. This day is taken as the Nobel prize ceremony day to honour the testator.

The Nobel Foundation was established under the terms of the Will of Alfred Bernhard Nobel, drawn up on the 27 November 1895, which in its relevant parts runs as follows: 1 'the whole of my remaining realizable estate shall be dealt with in the following way: the capital, invested in safe securities by my executors, shall constitute a fund, the interest on which shall be annually distributed in the form of prizes to those who, during the preceding year, shall have conferred the greatest benefit to mankind. The said interest shall be divided into five equal parts, which shall be apportioned as follows: one part to the person who shall have made the most important discovery or invention within the field of physics; one part to the person who shall have made the most important chemical discovery or improvement; one part to the person who shall have made the most important discovery within the domain of physiology or medicine; one part to the person who shall have produced in the field of literature the most outstanding work of an idealistic tendency; and one part to the person who shall have done the most or the best work for fraternity between nations, for the abolition or reduction of standing armies and for the holding and promotion of peace congresses. The prize for physics and chemistry shall be awarded by the (Royal) Swedish Academy of Sciences; that for physiological or medical works by the Karolinska Institute in Stockholm; that for literature by the (Swedish) Academy in Stockholm; and that for champions of peace by a committee of the persons to be elected by the Norwegian Storting (Parliament). It is my express wish that in awarding the prizes no consideration whatever shall be given to the nationality of the candidates, but that the most worthy shall receive the prize, whether he be a Scandinavian or not'.

Most probably, it was due to the influence of his secretary-cum-house-keeper that the Nobel prize for peace is one of the five Prizes that have been instituted, whereas the sixth Nobel prize for economic sciences has been established by the Swedish Riksbank since 1968. The other version about the Nobel prize for peace is that 'Evidence suggests that the award for peace may well have been the fruition of the inventor's long standing aversion to violence. Early in 1886, for example, he told a British acquaintance that he had a more and more earnest wish to see a rose red peace sprout in this explosive world'.

As stated above, the Nobel Prize for Physics and chemistry is awarded by the Swedish Academy of Sciences. In general, the Nobel Committee consisting of 3-5 members is elected for a period of three years for each Swedish Prize section by the Academy. (In later years, the number of members for the physics and chemistry groups has been fixed to 5 for each group.) The Nobel Committee sends out invitations during September to the competent persons to put forward proposals together with evidences. The persons who are eligible to submit proposals are:

1. Swedish and foreign members of the Academy of Sciences.
2. Members of the Nobel Committees for physics and chemistry.
3. Scientists who have been awarded the Prize by the Academy of Sciences.
4. Permanent and assistant professors in the sciences of physics and chemistry at the universities and institutes of technology of Sweden, Denmark, Finland, Iceland and Norway, and the Karolinska Institute.
5. Holders of corresponding chairs in at least six universities or university colleges selected by the Academy of Sciences with a view to ensuring the appropriate distribution of the commission over the different countries and their seats of learning.
6. Other scientists from whom the Academy may see fit to invite proposals. Decisions as to the selection of the teachers and scientists referred to in Expert's scrutiny and opinion of 3-5 members of the Nobel Committee shall be taken each year before the end of the month September.

SIR C.V. RAMAN

According to Nobel's will, the prize should have international character. Between the years 1901 and 1929; 28.6%, 20.0%, 20.0%, 8.6% and 8.6% Nobel Laureates belonged to Germany, France, England, USA and Scandinavian countries respectively; whereas the nominators who made proposals from the above countries were 25.9%, 13.9%, 7.6%, 11.0% and 13.4% respectively. As most of the nominators and proposed candidates belonged to these countries, the chances for others such as Russians were limited. On the other hand, these circumstances make the case study of Raman more interesting and highlighting.

In 1929, C. Fabry from Paris recommended J. Cabannes (Montpellier) and C. V. Raman (Calcutta), whereas N. Bohr proposed that either R. W. Wood or R. W. Wood and Raman should be considered for receiving the Nobel prize for physics. In that year 48 nominators sent 97 proposals and proposed in all 29 persons. Out of these 29 persons, L. de Broglie, Cabannes, Raman and Wood were declared by the Committee as the persons who fundamentally deserved the prize; but it was L. de Broglie who finally received the Prize for that year.

For the year 1930, 39 competent persons were asked to submit proposals. Out of them, 37 persons sent proposals. There were 21 valid recommendations for a full or shared Prize. Most of the recommendations were concerned with atomic theory and atomic physics. The atomic theory proposals had been worked out by Oseen. Out of the 21 nominations, Raman was the most suitable person; he was proposed 10 times, either as a single candidate for the Prize, or to share it with other physicists.

In that year some of the other scientists proposed included, M. Born, A. Sommerfeld, E. Schrodinger, W. Heisenberg, H. F. Osborn, and M. N. Saha (an Indian astrophysicist). The list was originally compiled for the physicists from the German-speaking area who had received the Prize before 1930. The list shows that Max von Laue (1879-1960) was the only physicist to have received unshared Nobel prize in a record time of two years after the discovery.

According to the Will of Nobel: 'The annual award of prizes shall be intended for works "during the preceding year" shall be understood in the sense that awards shall be made for the most recent achievements in the fields of culture referred to in the Will and for older works only if their significance has not become apparent until recently'.

The application and significance of the Raman effect becomes clear from the number of papers published within a period of one-and-half years after its discovery. 'By August 1929, Ganesan was able to compile a bibliography of 150 papers! The last date of sending the proposal was the first of February (in our case February 1930); and the list of original literature on the Raman effect, compiled by Kohlrausch, contained 225 entries till 31st January 1930, starting from the first publication of Raman and Krishnan in about the nature of the discovery.

Sir C.V. Raman's discovery found immediate applications in the field of experimental and theoretical physics; Raman was found to qualify for receiving the Prize in such a short period of time. The opinion of experts in the scientific community about Raman's scientific work, further substantiates his candidature for the prize.

The report prepared by the Nobel Committee was signed by H. Pleijel, Manne Siegbahn, V. Carlheim-Gyllensk"ld, Erik Hulthen and C. W. Oseen. Some amongst these were known personally to Raman. For example, M. Siegbahn (1886-1978), who received the Nobel prize in 1924 for his discovery and researches in the area of X-rays, had contacts with Raman. C. W. Oseen (1879-1944) who held the chair of theoretical physics at the University of Uppsala and later was appointed the director of the Nobel Institute for Theoretical Physics, was also known to Raman.

Niels Bohr (1885-1962), who received the Nobel prize in 1922 for investigation on the structure of atoms and the radiation emanating from them, had contacts with Raman. In a letter dated 21st March 1923 he wrote: 'I hope you will pardon the liberty I am taking in writing to you concerning Bidhubhusan Ray, who is one of the staff of this college and one of the promising young physicists of the Calcutta School. I have suggested to Ray that he might follow a different course (not to go to England and Germany like most Indians were doing) and that he cannot do better than spend a greater part of his period of deputation at Copenhagen working under your direction.' Raman visited Copenhagen, about which Bohr states, 'We often think of your visit here some years ago, and I hope very much that I shall have the pleasure of meeting you again before too long' (dated 18th September 1929). In the same letter he wrote, 'I take this opportunity to express my most cordial congratulations to you to your great discovery of the new radiation phenomenon which has added so immensely to our knowledge of optics and atomic physics'.

Raman wrote back to him, 'The great kindness you have shown me in the past encourages me to make a request of a personal character. As you know, my work on the new radiation effect has been received with enthusiasm in scientific circles, and I feel sure that if you give your influential support, the Nobel Committee for physics may recommend that the award for 1930 may go to India for the first time. The proposal for the award has to reach the Nobel Committee before 31st January 1930. I have greatly hesitated in writing to you about this, and it is only because I felt sure that you sympathise with the scientific aspirations of India that I have ventured to do so. With many apologies. I am, yours sincerely' (letter dated 6th December 1929).

Raman was not aware that he had been already nominated by Bohr for the Nobel prize in the year 1929. He repeated this decision for the year 1930. However, it was not the contacts alone, but rather it was his scientific contributions which fetched Raman the Prize.

The comments of some of the scientists who nominated Raman are:

C. T. R. Wilson (1869-1959), who received the Nobel prize in the year 1927, and E. Rutherford (1871-1937) wrote, 'There seems to be no doubt that a study of the change of frequency in liquid and solid media provides valuable information on the natural frequencies associated with the molecules–information which is difficult to obtain by other methods–and will prove of great service in increasing our knowledge. We are both of the opinion that Raman is a physicist of exceptional ability, who in the difficult conditions in his own country has built up a successful school of research which has already produced work of high quality. He is a man strong both on the theoretical and experimental side and this is well illustrated by many of his papers. We are of the opinion that his work is of the outstanding quality required for this great honour.' (Rutherford and Wilson to the Chairman, Nobel Committee, 25th January 1930).

J. Stark (1874-1957), who also got the Nobel prize in 1919 for his discovery of the Doppler effect in canal rays and the splitting of the spectral lines in an electric field, pointed out the practical side of the discovery by Raman. He observed, 'Answering your invitation I present to you a proposal for the Nobel prize in physics for the year 1930, and to be precise I propose: Professor C. V. Raman in Calcutta for the discovery of the effect of the change in the frequency of light when scattered which was named after him. This discovery means–independent of the transience of theories–a permanent progress in the knowledge of physical reality.' (J. Stark to the Nobel Committee, dated 7th January 1930. Translated from German.)

R. Pfeiffer from Breslau praised not only the discovery of the effect but also the earlier research works of Raman. He stated, 'I propose the professor of physics at the University of Calcutta Venkata Raman (Fellow of the Royal Society) for this year's Nobel prize for physics. Professor Raman has developed a fruitful research work since several decades from which I want to mention only his extensive investigations about the acoustics of Indian musical instruments and those about the diffraction of light in molecules (Tyndall phenomenon). These latter investigations led him to his great discovery namely the establishment of the effect which was named after him (Raman effect); as a result, Raman moved up to the very front of those physicists studying the problems of modern atomic physics. The Raman effect provides inner eigenfrequencies of the molecules that means a property of them due to their inner constitution, that means it (the effect) provides a powerful method for the exploration of molecules. It is of particular interest that the eigenfrequencies determined by this method which are situated in the spectral range of ultra-red frequently cannot be found by the help of spectroscopic methods. Therefore ultra-red research and Raman effect have a stimulating influence upon one another so that everyday surprises us with fresh evidence. In summary, I have to express my conviction that the Raman effect is one of the most important and most fruitful discoveries of the last years.' (R. Pfeiffer to the Nobel Committee, 22nd January 1930.

Niels Bohr (1885-1962) stated, 'This phenomenon (Raman effect), the explanation of which agrees so well with the quantum theoretical ideas, will undoubtedly become a most important source in increasing our knowledge of the states of the atoms or molecules of matter in transitions, between which their characteristic spectra are emitted.' (N. Bohr to the Nobel Committee for physics, 29th January 1929).

The above evidences amply show that experts in the field recognized his work, which qualified him for the Nobel prize.

In his book Raman and His Effect, Keswani wrote, 'Raman was a great teacher and beloved of his pupils but he could have given more credit to K. S. Krishnan who contributed no less to the discovery of the effect now known as the Raman effect'; and further, 'Many have felt that K.S. Krishnan should have been acknowledged as the co-discoverer of the effect now bearing Raman's name exclusively'. As far as the coining of the term Raman effect (in English and German-speaking areas) or Smekal-'Raman effect' is concerned, the term was coined by scientists other than Raman. In the same book Keswani has raised some interesting questions, such as 'Why did the Nobel Committee for physics not vote for the sharing of the prize by the Indian(s) and Russians?'.

The names of the collaborators were known to the Committee as well as to the expert who prepared the report. The name of candidates given in the report of the Nobel Committee shows that not a single collaborator of Raman was voted by the nominators. It was not possible for Raman to put his name or any of his co-workers for the nomination. The prize was awarded not only for the discovery but also for 'his work on light scattering and the discovery of the Raman effect'. The committee had to base their decision on the proposals and the opinions of the experts. If there was a controversy among the members of the Committee regarding awarding the prize to the co-workers as well, it will never come to light because under the rules of SNF 10, 'Proposals received for the award of a prize, and investigations and opinions concerning the award of a prize may not be divulged. Should divergent opinions have been expressed in connection with the decision of a prize-awarding body concerning the award of prize, these may not be included in the record or otherwise divulged'.

There existed a controversy between Raman and his Russian colleagues on the priority of the discovery. Raman wrote in 'Nature', `The Russian physicists, to whose observation on the effect in quartz about which Darwin refers, made their first communication on the subject after the publication of the notes in 'Nature' of 31st May and 27th April. Their papers appeared in print after sixteen other printed on the effect, by various authors, had appeared in recognized scientific periodicals.'

Since the controversy was a public knowledge, the Committee had to pay special attention to this issue despite the fact that the Russian scientists were recommended only twice. Papalexis of Leningrad (now St. Petersburg) proposed that the prize should be awarded to Mandelstam (1879-1944) alone, whereas Chwolson stated that Raman should get half of it and the rest of it should go to Landsberg (1890-1957) and Mandelstam.

The arguments put forward by Chwolson from Leningrad in his proposal follows: 'that Raman shares the honour of his discovery with Landsberg and Mandelstam; because undoubtedly, they discovered the named phenomenon in quartz crystals at the same time and independently from Raman. Only due to the external circumstances and the negligence of the Russian researchers in publishing their discovery, is according to Chwolson, the cause that the effect has been named as the "Raman effect". Raman's discovery was made on February 16th and Landsberg and Mandelstam's on 21st February 1928. Letters by Raman and Krishnan to 'Nature' were published on 31st March 1928, whereas Landsberg and Mandelstam's first publication in Naturwissenschaften on 13th July. By giving Raman alone the prize the Russian scientists would be severely punished due to this negligence', said Chwolson. Also he gives the reference of M. Born from Goettingen, who was said to know the exact situation and to have given his opinion publicly. Chwolson obviously refers to Born's article in Naturwissenschaften (1928, 16, 741) on the Fourth Russian Physicists Conference, where Born was a guest. Born mentions in his lecture that the phenomenon was discovered in Moscow and Calcutta at the same time and the Russian physicists should share the honour with Raman'.

About the proposal of Papalexis from Leningrad the Committee observed, 'An even stronger impression of this opinion we find in the proposal of Papalexi, Leningrad in which he writes that Mandelstam alone should get the prize. He supports his proposal as follows: Mandelstam has been working since 1907 on the theoretical and experimental aspects of the diffusion light. Since 1918 through his theoretical interpretations he came to the idea of the existence of the scattering light, which corresponds to the Raman effect. Papalexi refers to an article of Mandelstam in Journal of Russian Physical Society (Journal d. russ. phys. Gesellschaft) 1926, The Committee was of the opinion that Smekal in 1923 and Kramers and Heisenberg in 1925 had already given this explanation.

In order to make the position of the Committee on this matter clear, it will not be out of place to quote the comments of it, which stated, 'If we see Mandelstam's and Landsberg's first publication in Naturwissenschaften (1928, 16, 557), we get a different picture. The short note had been dated 6th May and explains the discovery of combination lines of the diffuse light in crystalline quartz. The existence of these lines has been shown experimentally, but about the interpretation of the lines the authors say, "We consider it to be premature at this moment to give a definite interpretation of the phenomenon in question– one of the theoretical interpretations, the authors gave the same statement as Raman. And further they said, "Whether and in what way the phenomenon observed by us is connected with the one which was recently described by Raman cannot be judged at the moment because of its rather summary description".

However, Raman's and Krishnan's letters of 31st March as well as that of 21st April gave a very clear explanation of the nature of the phenomenon, (both cited by Mandelstam and Landsberg). Under these conditions, Mandelstam and Landsberg cannot argue to have obtained their experimental results independently.

According to experts, 'A work may not be awarded a prize, unless it by experience or expert scrutiny has been found to be of such outstanding importance as is manifestly intended by the Will'. The expert who gave this report was Erik Hulthen, professor and the director of the Physical Institute and the member of the Nobel Committee from 1929 to 1962. He prepared the report on the Raman effect under the title 'The complete explanation of the Raman's-Effect'. The report was in favour of Raman and cited work done by different scientists on the topic in the past years. The conclusion of the report follows:

1. The proposal by Chwolson that the Nobel prize should be divided between Raman and Landesberg-Mandelstam had been rejected because they did not come to an independent interpretation of their discovery.
2. For the same reasons, the proposal of Papalexis in favour of Mandelstam had not been taken into account.
3. The uncertainties concerning the explanation of the intensity of Raman and infrared lines in the spectrum, could be explained during the last year.

4. The Raman method has been applied with great success in different fields of molecular physics.

5. The Raman effect has effectively helped to check the actual problems of the symmetry properties of molecules, thus the problems concerning the nuclear-spin in the atomic physics'.

The Nobel Committee said, 'the Raman effect is useful for the study of atomic physics and the constituents of compound. It also gives valuable information to prove modern theories in atomic physics. The Committee finds Raman's discovery on diffusion light is worth the Nobel prize for physics.

We have seen that the expert as well as the Nobel Committee had given their award-adjudication in favour of Raman. Thus he cannot be held responsible for not sharing the prize.

According to the regulations, the Nobel Committee submitted its report with proposal and opinion to the Swedish Academy of Sciences on 20th September 1930 (the regulation says-by the end of September) with the conclusion, 'The Committee has decided to ask the Academy to award the Nobel prize for physics for the year 1930 to Chandrasekhra Venkata Raman, Calcutta, for his work on 'the diffusion of light' and for the effect named after him'.

According to the regulations, 'the Committee shall take up the matter for a final decision before the middle of the following 27th of November. Another committee, 'Physic-class' comprising about 25 members is responsible for controlling all the proposals, recommendations and other documents concerning the Nobel prize for physics. It gives its opinion and final decision to the Academy about the selection of the Nobel Laureate. According to the Special rules of SNF 7, 'The Academy shall take the matter for a final decision before the middle of the following November'. 'The laureate is immediately notified of the decisions, which are then announced internationally at a press conference held in Stockholm and attended by representatives of the international news media. The messages contain the names of the laureates and a short statement describing the reason for award'. Raman would not have been in time to attend the Nobel prize giving ceremony on 10th December 1930, if he had not been informed by proper channels. Two months before the awarding, Raman knew that he was awarded the Nobel prize. He had the supreme audacity of booking his steamer passage to be in time for the ceremony at Stockholm. That not only did he take such a step but went further and declared publicly that he did so are both interesting facts of his life. Whatever might be the source of his information, on 10th December 1930 he and Lady Raman were in Stockholm to receive the prize.

According to SNF rules, 'It shall be incumbent on a prize-winner, whenever this is possible, to give a lecture on a subject relevant to the work for which the prize has been awarded. Such lecture should be given within six months of the Festival Day in Stockholm, or, in the case of the Peace Prize, in Oslo.' Raman delivered his Nobel lecture on 11th December 1930, entitled `The Molecular Scattering of Light', in which he gave the point of motivation of his research, i.e. the blue colour of the sea and further extension of this work on the scattering of light, which led to the discovery of the effect.

After receiving the prize Raman stayed on for further five days. On 17th December 1930 he was in the Grand Hotel Olso from there he wrote a letter to Bohr which dictates, ' We next proceed to Goeteborg and then to Copenhagen which we reach on the night of the 19th December or perhaps the 20th December. It is my hope to (be) able to meet you and your group of investigators and to spend three or four days in Copenhagen much to the advantage of my knowledge of physics' (dated 17th

December 1930). The telegram which he sent from Goeteborg shows that in Copenhagen he delivered a lecture on 'Scattering of light in crystals'. Raman met A. Sommerfeld in Munich. After this tour through Europe he returned to India.

Thus, Raman received the Nobel prize 'for his work on diffusion of light and for the effect named after him'. The objections raised by some historians that Raman did not share the Nobel prize with others or that the Committee ignored Raman's collaborators as well as Russian colleagues is not correct; as he was awarded the Prize not only for the Raman effect, but for other work in this field as well. The Nobel Committee had to take the decision according to certain rules and regulations imposed on it by the Nobel Foundation. Raman was nominated 10 times and the nominators wrote convincing recommendations in favour of him; thus the Committee decided for Raman. He received the Nobel prize in record time due the practical significance of the discovery, as well as the good opinion of the famous contemporary scientists about his work.

SUBRAHMANYAN CHANDRASEKHAR

In 1946, a boyish-looking man was driving once every week from Yerkes Observatory in Wisconsin to the University of Chicago, a distance of about 160 kilometers to teach two students. This teacher was Subrahmanyan Chandrasekhar, one of the world's leading astrophysicists. Chandrasekhar believed in the theory-'great men are seldom born. They are self made'.

S. Chandrasekhar was born on October 19, 1910 in Lahore, then part of undivided India, now in Pakistan as the first son and the third child of a family of four sons and six daughters. His early education, till he was twelve, was at home by his parents and by private tuition. His father, Chandrasekhara Subrahmanya Ayyar, an officer in Government Service in the Indian Audits and Accounts Department, was then in Lahore as the Deputy Auditor General of the Northwestern Railways. His mother, Sita was a woman of high intellectual attainments (she translated into Tamil, for example, Henrik Ibsen's A Doll House), was passionately devoted to her children, and was intensely ambitious for them. In 1918, his father was transferred to Madras where the family was permanently established at that time.

He continued his basic education in Madras, and was into reading from a very young age. While most of his class mates never read books outside the syllabus, he visited the library regularly and read every new book in Physics, even research journals!

Those were the 1920's when modern physics was taking birth and there was a flood of new, exciting discoveries. In particular, books of great scientists like Arnold Sommerfield and Arthur Crompton drew Chandrasekar to modern physics. And before he was 18, his research papers started appearing in the 'Indian Journal of Physics'.

Meanwhile, he attended the Hindu High School, Triplicane, during the years 1922-25. His university education (1925-30) was at the Presidency College. He took his bachelor's degree, B.A. (Hon.), in physics in June 1930 and by that time he had many papers to his credit. One of them even appeared in the proceedings of the royal society, a rare honour for one so young. In July of that year, he was awarded a Government of India scholarship for graduate studies in Cambridge, England. In Cambridge, he became a research student under the supervision of Professor R.H. Fowler (who was also responsible for his admission into Trinity College). On the advice of Professor P.A.M. Dirac, he spent the third of his three undergraduate years at the 'Institute of Teoretisk Fysik' in Copenhagen. At the age of 27, Chandrasekhar's reputation as a promising astrophysicist has been firmly established.

He took his Ph.D. degree at Cambridge in the summer of 1933. In the following October, he was elected to a Prize Fellowship at Trinity College for the period 1933-37. During his Fellowship years at Trinity, he formed lasting friendships with several scientists, including Sir Arthur Eddington and Professor E.A. Milne. While on a short visit to Harvard University (in Cambridge, Massachusetts), at the invitation of the then Director, Dr. Harlow Shapley, during the winter months (January-March) of 1936, he was offered a position as a Research Associate at the University of Chicago by Dr. Otto Struve and President Robert Maynard Hutchins. He joined the faculty of the University of Chicago in January 1937.

During his last two years (1928-30) at the Presidency College in Madras, he formed a friendship with Lalitha Doraiswamy, one year his junior. This friendship matured; and they were married (in India) in September 1936 prior to his joining the University of Chicago. In the sharing of their lives during the next forty-seven years, Lalitha's patient understanding, support, and encouragement were the central facts of his life.

After the early preparatory years, his scientific work followed a certain pattern motivated, principally, by a quest after perspectives. In practice, this quest consisted of his choosing (after some trials and tribulations) a certain area which appears amenable to cultivation and compatible with his taste, abilities, and temperament. And when after some years of study, he felt that he has accumulated a sufficient body of knowledge and achieved a view of his own, he went on to present his point of view, in a coherent account with order, form, and structure.

There have been seven such periods in his life: stellar structure, including the theory of white dwarfs (1929-1939); stellar dynamics, including the theory of Brownian motion (1938-1943); the theory of radiative transfer, including the theory of stellar atmospheres and the quantum theory of the negative ion of hydrogen and the theory of planetary atmospheres, including the theory of the illumination and the polarization of the sunlit sky (1943-1950); hydrodynamic and hydromagnetic stability, including the theory of the Rayleigh-Bernard convection (1952-1961); the equilibrium and the stability of ellipsoidal figures of equilibrium, partly in collaboration with Norman R. Lebovitz (1961-1968); the general theory of relativity and relativistic astrophysics (1962-1971); and the mathematical theory of black holes (1974-1983).

In the world of astrophysics, Chandrasekhar is best known for 'Chandrasekhar limit'. This imposes a restriction on the size of a highly dense variety of star known as the 'White Dwarf'. If this type of star has mass in excess of limit, it explodes like thousands of nuclear bombs ignited together to become a very bright star called the 'Supernova', until all excess matter is shed into space. Although Chandrasekhar calculated his 'limit' purely on mathematical equations, astronomers have found that all the 'White Dwarf' stars in the sky have masses within his prescribed limit.

As early as 1935, he had come close to speculating on the formation of 'Black Holes', super heavy heavenly objects, a spoonful of which weighs several thousands of tons. But contemporary astronomers were not prepared to accept that intriguing speculation.

Chandrasekhar also made significant contributions to understanding the atmosphere of stars and the way matter and motion are distributed among stars in a galaxy. His work on rotating fluid masses and the blueness of the sky is also well known. For his contributions to the study of the stars, Chandrasekhar received the highest award in science, the Nobel Prize in Physics in 1983.

NASA's premier X-ray observatory was named the 'Chandra X-ray Observatory' in honour of the Indian-American Nobel laureate, known to the world as Chandra (which means "moon" or "luminous" in Sanskrit), he was widely regarded as one of the foremost astrophysicists of the twentieth century. Chandra immigrated in 1937 from India to the United States, where he joined the faculty of the University of Chicago, a position he remained at until his death. He and his wife became American citizens in 1953.

Trained as a physicist at Presidency College, in Madras, India and at the University of Cambridge, in England, he was one of the first scientists to combine the disciplines of physics and astronomy. Early in his career he demonstrated that there is an upper limit - now called the Chandrasekhar limit - to the mass of a white dwarf star. A white dwarf is the last stage in the evolution of a star such as the Sun. When the nuclear energy source in the centre of a star such as the Sun is exhausted, it collapses to form a white dwarf. This discovery is basic to much of modern astrophysics, since it shows that stars much more massive than the Sun must either explode or form black holes.

Chandra was a popular teacher who guided over fifty students to their Ph.Ds. His research explored nearly all branches of theoretical astrophysics and he published ten books, each covering a different topic, including one on the relationship between art and science. For 19 years, he served as editor of the Astrophysical Journal and turned it into a world-class publication. In 1983, Chandra was awarded the Nobel prize for his theoretical studies of the physical processes important to the structure and evolution of stars. According to Nobel Laureate Hans Bethe, "Chandra was a first-rate astrophysicist and a beautiful and warm human being. I am happy to have known him." "Chandra probably thought longer and deeper about our universe than anyone since Einstein," said Martin Rees, Great Britain's Astronomer Royal.

Recipient of many medals and honours, Chandrasekhar was at once a physicist, an astrophysicist, and an allied mathematician. Every decade or so, he will change his field of study. He would study a new subject from scratch, master it and make significant contributions to it.

Though he lived in USA since 1937, a visitor to his house might well have found him wearing the south Indian 'dothi', and listening to 'carnatic music'. The world lost this great soul on August 21, 1995.

Professor Chandrasekhar was one of those rare scientists who lived the life of a scientist from their youth till they breathed their last. He published his first scientific paper entitled "Thermodynamics of the Compton Effect with Reference to the Interior of the Stars" in the Indian Journal of Physics when he was eighteen years old. His last published work, "Newton's 'Principia' for the Common Reader" appeared in 1995, the year he died. He was eighty-four years old then. His professional output during his long scientific career, spanning nearly seventy years, was phenomenal. For his outstanding contributions to astrophysics and mathematical physics he received a large number of awards from professional societies of physics and astrophysics. Universities conferred upon him their honorary doctorate degree. The Presidents and the Heads of States felt honoured in honouring him. But he valued more the professional recognition from his peers and students. On his 73rd birthday in 1983 the Nobel Prize for Physics was announced for

Chandrasekhar. Recently, the NASA has named its next satellite for scientific research "Chandra", the name by which his colleagues and admirers called him.

Chandrasekhar's life and work have been viewed with varied perspectives. Many books on Chandrasekhar have been written. There are two inspiring biographies of Chandrasekhar – One is the biography by Kameshwar C. Wali, entitled "Chandra". The other is the book, "Chandrasekhar and His Limit", written by G. Venkataraman. The author himself paid a tribute to Chandrasekhar in an article published in the University News in 1996. The title of that article is 'S. Chandrasekhar - As I Knew Him'. It gives the perspective of a pupil of his relationship with his teacher. In another article entitled "Subrahmanyan Chandrasekhar: A Brilliant Star Without Limit" the author has worked out the calculation that Chandrasekhar at the age of 20 years might have carried out during his sea voyage to England. In that article an analysis of the concepts that Chandrasekhar made use of in working out the fundamental mass, known as the 'Chandrasekhar limit', has been given. Therefore, *raison of d'être*, a new article on Chandrasekhar can at best be the presentation of a perspective different from that of the articles already published by the author. As this article is being written for a special issue of the Indian Journal of Mathematics Education being brought out under the aegis of the Delhi Mathematics Teachers' Association the emphasis is on reasoning and not on the advanced mathematics used by Chandrasekhar in carrying out his calculations.

Chandrasekhar is best known for his path breaking work on the physics of white dwarfs. He could explain how the stars would settle down after their burn out. He made brilliant use of the newly discovered special theory of relativity and the quantum statistical mechanics in the classical model of structure of stars and arrived at an expression of a critical stellar mass. In the following an attempt has been made to bring out the line of reasoning that led Chandrasekhar to reach his conclusion on the fate of stars. Besides making significant contributions to the study of the stars, he has written books that are recognised as classics.

There are innumerable numbers of stars in the universe. In our own galaxy the 'Milky Way' it is estimated the number of stars is of the order of hundred thousand million. The Sun is a typical star of the Milky Way. It is located at a distance of about 30,000 light years from the centre of the galaxy. We at the Earth owe all the life that exits on our planet to the Sun. The Earth along with the other planets of the solar system was created out of the same debris of gas and dust that formed the Sun. It is, therefore, reasonable to hypothesis that the laws of physics that have been discovered by performing terrestrial experiments will also be applicable to the Sun, the other stars in the galaxy and even the universe itself.

In the twentieth century new types of astronomy such as radio, infra red, X-ray, gamma ray and the satellite-based telescopes became available as windows to view the universe. New astrophysical objects such as the pulsar, the black hole and the cosmic microwave radiation were discovered and interest increased in the study of stars, galaxies and the universe using the laws of physics. The special and the general theory of relativity, the quantum mechanics and the nuclear physics emerged as the new physics in the first half of the twentieth century. It was realised that though the new physics was discovered through the study of atomic and nuclear phenomena it is likely to be equally important in understanding the large scale structures such as the stars.

The pioneering work of Chandrasekhar was carried out during the period 1930 - 1935. He used the new physics, also known as the modern physics, in determining theoretically how the stars find their peace or in other words what happens when the stars die. As a by product of his work he could anticipate the existence of black holes forty years before these esoteric objects were discovered.

Astronomy has made it possible to see stars in different stages of their life cycle. Like the life we know on the Earth, each star is born, it evolves with time and ultimately dies either undergoing a catastrophic phase such as the supernova explosion or without it and settles down as a white dwarf or as a pulsar or as a black hole. The end of a star was expected to be decided by its mass, as all other properties such as its initial chemical composition would have been obliterated by the time the fusion process, the source of energy production, stops. As already mentioned Chandrasekhar found a critical mass called the 'Chandrasekhar limit', a benchmark that determines the ultimate fate of a star. The numerical magnitude of the 'Chandrasekhar limit' is expressed in the unit of mass of the Sun. In this unit it is equal to 1.4 times the solar mass.

Stars have masses varying from a fraction of the solar mass to hundreds of solar mass. When we see the night sky we see stars of different masses and of different ages. Measurable quantities that describe the state of a star are its absolute luminosity and its surface temperature. Hertzsprung and Russell made a scatter plot of absolute luminosity and temperature of stars. They saw that most of the stars in their graph, the H-R diagram, lie in a narrow band. This band is called the main sequence. It was noticed that some of the stars lie in a corner of the H-R diagram away from the main sequence. These stars are the dwarfs, as the estimate of their radius show that in size they are a fraction of the size of the Sun. Typically the size of white dwarfs is of the order of that of the Earth. Their mass is about that of the Sun. Therefore the white dwarf are highly dense objects that were stars before they reached this stage. Prior to the work of Chandrasekhar it was generally believed that all stars when they exhaust their nuclear fuel would become white dwarfs. The astrophysicists speculated that after a star has burnt out its nuclear fuel some new physics might help it in settling down as a white dwarf. In this case the new physics was the degeneracy pressure exerted by the electron gas. Unlike the thermodynamic pressure due to gas and radiation that provide equilibrium to a star against the inward gravitational pull as long as a star produces energy by burning its nuclear fuel, the degeneracy pressure does not require energy production. It is a quantum mechanical effect.

By the first quarter of the twentieth century astrophysicists, most well known of whom was Eddington, had worked out the theory of stellar structure using the Newtonian theory of gravitation, the equation of state of a polytrope and the equations of hydrostatic equilibrium. Using his model Eddington could explain the main sequence band in the H-R diagram. But the Eddington's model could not explain the white dwarfs as their positions in the H-R diagram were outside this band. This situation became a pointer to the need of invoking some new physics for going beyond the limitations of the classical model.

The genius of Chandrasekhar was that using the special theory of relativity and the quantum statistical mechanics as keys, he explored the physics of the white dwarf and discovered new features. He had learnt the quantum statistical mechanics required for working out the equation of state of an electron gas from the lectures given by Arnold Sommerfeld in Madras in 1928. The electrons are Fermions; fundamental particles with spin 1/2. They cannot occupy the same state. When electrons are compressed they resist being squeezed and provide a pressure called the degeneracy pressure. The electron gas will therefore obey the Fermi - Dirac statistics. His estimate of the velocity of electrons inside a white dwarf indicated that the electron gas could be relativistic. Therefore the pressure of the electron gas may have to be worked out using the Fermi - Dirac statistics in the relativistic limit. He used the theoretical model for white dwarf stars developed by Fowler, the person under whom he later did his Ph.D. research. But he used in his calculations the equation of state for a relativistic degenerate electron gas instead of the non-relativistic case considered by Fowler. The details of this calculation can be seen from the author's article and also from Chandrasekhar's book "Introduction to the Theory of Stellar Structure."

Chandrasekhar found that if a star has a mass less than the critical mass it would find peace as a white dwarf. But if the star has a mass greater than the critical mass it will become unstable against the inward gravitational pull and it will collapse. The possibility that an object that initially had a mass and size more than that of the Sun will shrink in size and disappear as a point was revolutionary and mind boggling. His startling findings were ridiculed by no less a person than Eddington. The fall out of this controversy was that for nearly two decades scientific interest in the physics of massive astrophysical objects remained dormant.

Chandrasekhar pursued the physics of black hole in the last quarter of his scientific career. He published his magnum opus "Mathematical Theory of Black Holes" in 1983. An elementary introduction to black holes has been given by the author in his article "Beyond White Dwarfs, Toward Black Holes - An Introduction to S. Chandrasekhar."

To give a brief on the non-mathematical account of the fundamental work carried out by Chandrasekhar in the initial years of his scientific career, it is difficult to go further without bringing in the mathematics.

The morning James Cronin, University of Chicago Professor Emeritus in Physics and Astronomy and Astrophysics, won the

1980 Nobel Prize in physics, he was temporarily unavailable to speak with journalists because he wanted to attend a class that Subrahmanyan Chandrasekhar was giving on general relativity.

Chandra received the 1983 Nobel Prize in physics. Chandra's uncle, C.V. Raman, received the 1930 Nobel Prize in physics for discovering the Raman effect, which describes the diffraction of light by crystals. Raman was the first Asian to receive a Nobel Prize in science.

Chandra's commitment to teaching was legendary. In the 1940s, he drove 200 miles round trip each week from Yerkes Observatory in Williams Bay, Wisc., to the University to teach a class on stellar atmospheres. One day he insisted on driving from Yerkes to teach the class despite a heavy snowstorm. Chandra ended up teaching a class of only two that day. The two students—Tsung Dao Lee and Chen Ning Yang—won the 1957 Nobel Prize in physics, obtaining the distinction even before their professor.

In 1930, at the age of 19, Chandra completed his degree at Presidency College in Madras, India, and boarded a boat to England for graduate study at Cambridge University. While on the voyage, Chandra developed a theory about the nature of stars for which he would be awarded the Nobel Prize 53 years later. His theory challenged the common scientific notion of the 1930s that all stars, after burning up their fuel, became faint, planet-sized remnants known as white dwarfs. He determined that stars with a mass greater than 1.4 times that of the sun—now known as the Chandrasekhar limit—must eventually collapse past the stage of white dwarf into an object of such enormous density that "one. is left speculating on other possibilities," he wrote.

Chandra's Nobel Prize-winning theory initially was rejected by peers and journals in England. The distinguished astronomer

Sir Arthur Eddington publicly ridiculed Chandra's suggestion that stars would collapse into such objects, which are now known as black holes. Disappointed and reluctant to engage in public debate, Chandra moved to the United States in 1937 and joined the University of Chicago faculty. Today, the extremely dense neutron stars and black holes implied by Chandra's early work are a central part of astrophysics.

Chandra's paper about the mass limit of white dwarf stars initially was rejected by the Astrophysical Journal as scientifically unsound. After receiving detailed proof of the formula on which the paper was based, the journal published his paper. Chandra served as managing editor of the Astrophysical Journal from 1953 until 1971.

Chandra was widely known for his appreciation of literature, music and the philosophy of science. The depth of his knowledge in these areas was evident in a book on his Philosophy of Aesthetics in Science, Truth and Beauty: Aesthetics and Motivations in Science, and in his frequent lectures on the relationship between the arts and the sciences.

Chandra developed his own astonishing style of research that entailed tackling first one field of astrophysics and then another in great depth. He wrote definitive books describing the results of his investigations on topics ranging from radiative transfer of energy through the atmospheres of stars to the motions of stars within galaxies, and from magnetohydrodynamics to Einstein's theory of general relativity and black holes.

Chandra received 20 honorary degrees, was elected to 21 learned societies and received numerous awards in addition to the Nobel Prize, including the Gold Medal of the Royal Astronomical Society of London; the Rumford Medal of the American Academy of Arts and Sciences; the Royal Medal of the Royal Society, London; the National Medal of Science; and the Henry Draper Medal of Science.

A student's (Ms.Maheswari) tribute to Prof.S.Chandrasekar.

My association with Professor Chandrasekhar dates back to 1964 when I reached the University of Chicago to do my Ph.D. studies in physics. I saw an Indian looking Professor dressed in a black suit wearing a Cambridge University tie. He was sitting in the first row of a physics colloquium and I could easily connect that the distinguished person was Professor Chandrasekhar, whom everybody affectionately called Chandra. He appeared to me then both very reserved and unapproachable. I changed this opinion as I began to know him more closely. My next encounter with Chandra was at a Physics-Club meeting, which were specially arranged by the Department of Physics of the University to provide to the fresh class of graduate students face-to-face interaction with the senior faculty of the University. Professor Chandrasekhar spoke on General Theory of Relativity and its relevance to Cosmology and Astrophysics. I do not think I followed the lecture, but can distinctly recall the remark made by Chandra, "Veracity of the Einstein's theory of Gravitation is as undisputable as are the findings of Justice Warren on the assassination of President John F. Kennedy". I left the meeting with more awe and a feeling of a vast distance between his intellectual abilities and what I possessed as a twenty-one year old graduate student.

I was thrilled to see the announcement that Professor Chandrasekhar would teach a course on Non-relativistic quantum mechanics. I had studied this course as a part of M.Sc. studies in physics at the University of Delhi. I signed up for that course perhaps thinking that I should be able to impress the Professor with my background and the head start I thought I had. In the first lecture when the Professor entered the class he demanded that no smoking be observed, as he was allergic to tobacco smoke. This was obeyed by the class, but Chandra's reasons were suspect as he could be seen sitting between Professor Mark Ingraham and Professor Gregor Wentzel, both of whom puffed away cigar smoke continuously and within their vicinity the pollution level could only be matched by what comes out of coal-fed boilers in Chicago City. Chandra spoke Cambridge-English without a trace of American accent and wrote on blackboard as though he was doing calligraphy. He did not like being disturbed during his lecture and looked reprovingly at students drinking coffee or eating sandwiches. The course was uneventful as it progressed but a jolt was experienced by the entire class when he announced at the end of the eleventh week that the examination would be of six hours duration with an optimum response time of about four hours. He further elaborated that there would be only one problem to be solved in closed-book/closed-notes setting and that rough-calculation sheets were to be appended to the answer-script. He also advised students to bring their pack lunch to the examination hall! The problem to be solved turned out to be on finding analytically, changes in energy levels of hydrogen atom in strong electric field by setting up the Hamiltonian and writing the Schrodinger equation. Hints were given for various stages of solution that could be reached after about each successive hour of work. I vaguely recall that I could not proceed further beyond the fourth hour and closed my test after eating sandwiches, which I had specially prepared and rounded up my snack by an apple. I wrote from National Radio Astronomy Observatory, Green Bank, where I had gone to undergo summer training, with misgivings, to the Graduate Students Advisor of the Department of Physics to let me know how I had fared in Professor Chandrasekhar's course and whether the University would continue to give me financial assistance in the next academic year. I received a reassuring reply that Professor Chandrasekhar was happy with my performance and that the University would be pleased to support my further graduate studies. I have narrated this incident at length because it brings out how a teacher probed the mind of each of his students so painstakingly and without complaining of an inordinate demand on his time in spite of pressure of research and other professional commitments.

Although, I chose to work in the field of theoretical high-energy physics, not of direct research interest of Professor Chandrasekhar, Chandra decided to be one of the four advisors for supervising my doctoral studies. I began to experience Chandra's warmth from smile on his face in acknowledging my greetings. Gradually I started to know the real Chandra and Mrs. Chandrasekhar. Occasionally I would join both of them at dinner table in the restaurant of the International House and listen to episodes from the life of the esteemed Professor, as narrated by his wife. She knew how uncomfortable students were in Chandra's presence and that we felt elated and inspired on being chosen to be shared stories from the life of the great man. One story on how Chandra handled his graduate students I contribute to this essay, because I was the second party in the incident. I wanted to fix up with my advisory group the date and time for holding an assessment, a requirement of the Ph.D. course. Meeting Chandra in his Office in the Laboratory for Space Research and Astrophysics was difficult as an appointment was required. But I knew Chandra's daily habits and decided to catch him during his walk to the laboratory. I accosted Chandra and asked point blank whether he would be in station on such and such date and whether he could be available for conducting my assessment. He suggested that I defer the assessment for a week. I told him, 'Chandra, do I not come under your priority and can you not spare half-an-hour for a graduate student?' Chandra's immediate response was a yes to the scheduling of my assessment on the date I had proposed but he said, 'Maheshwari, can you explain the concept of negative temperature?' Chandra continued to remind me whenever I met him since that I had pleaded to him to postpone the assessment for another month so I could prepare myself better.

In between I used to meet Chandra to discuss physics and sometimes he would walk to me at my desk with some newspaper reports on India in his hand and share his anguish. He once asked me to explain to him the concept of pseudo energy-momentum tensor for the gravitational field. I felt honoured in having been approached by the Professor but specially privileged when Professor Chandrasekhar gave me a person-to-person seminar on how he had used this concept in his research work. This aspect of his life is also important because he took pride in pointing out that he benefited in research more from his students than from his colleagues in the University. In 1969 he told me that during the course of his career in the University except for one research paper, which he had jointly written with Enrico Fermi, all his research work was either independent or was carried out with his graduate students.

He would emphasise to me the importance of diligence and observance of discipline in daily working habits. He emphasised that personal targets had to be continually advanced further so that life may remain an unending challenge without ever getting the feeling of having arrived at.

He once mentioned that in having decided to live abroad he could only live the life of a scientist. From his own experience he pointed out that living the life of a scientist in a foreign country is extremely difficult and very rarely and very few persons can hope to contribute to science at levels that bring lasting recognition and scientific immortality. At the age of nineteen,

Chandrasekhar had made the scientific discovery of the existence of a fundamental stellar mass from his study of the physics of white dwarf stars, the famous Chandrasekhar limit. Although Chandrasekhar had carried out his monumental work during his long sea voyage to England from India in 1930 and published it in 1931 in the Astrophysical Journal of the University of Chicago, but was awarded the Nobel Prize in physics for this work only in 1983. Chandra did not let recognitions slow down his pace of work and kept on moving his targets throughout his life; to wit physics of white dwarf, stellar structure and radiative transfer, magneto-hydrodynamics, mathematical theory of black holes, study of Newton's Principia.

He was a perfect embodiment of what he practised and his advice to his students was based on his experience. He might have influenced me in deciding to return to India after getting my Ph.D. degree. In what follows next I would describe the role he played in my later professional life.

Professor Chandrasekhar was happy to know when I informed him that I had joined the University at Simla. Once, he wrote to me that while taking a walk at Aspen in the Rocky Mountains in the U.S.A. he imagined that in Simla I would also be similarly situated in an ideal setting conducive for pursuing theoretical physics. Soon after, in early 1973, I met Chandra in New Delhi. He asked me how I was progressing with my work and if there was something he could do to help me. I told Chandra, "Nice climate and beautiful natural environment are fine but I need journals to do my scientific work, which the new University I had joined was unable to provide me." Chandra on the spot decided to gift to the Himachal Pradesh University his entire personal collection of journals. Within three months of that fateful meeting the Himachal Pradesh University received collection dating back to 1935 of the Physical Review, the Physical Review Letters and the Reviews of Modern Physics. This gift by Professor Chandrasekhar was without any expectation in return except that the journals should be made available for research consultation to all students and faculty. This act of generosity is unparalleled and brings out his genuine concern for his students and interest in their academic growth. What has been described here is a humble tribute of an Eklavya to his Dronacharya. His other pupils will have similar stories to recount on how this great teacher influenced their life.

ORNITHOLOGIST

Dr. SALIM ALI

Most of us like watching colourful birds but there was one man who was so involved in their study that he was nicknamed 'The Birdman of India'. This man also known as the 'Grand Old Man of Indian Ornithology' is Salim Ali.

Born on November 12, 1896, orphaned at a very young age, Salim Moizuddin Abdul Ali was brought up by his maternal uncle. He spent his childhood in a large house, which he described in his autobiography as being "full of a miscellaneous assortment of other orphans and children of absentee friends and relations of different ages."

As a child, Salim was given an expensive Daisy air gun as a present, and spent all his time shooting sparrows around the house. One day he noticed that one of the sparrows he had shot had a yellow throat. He couldn't hold his curiosity and approached his uncle. His uncle who was equally clueless took him to the Bombay Natural History Society (BNHS) in the hope of finding an answer. There, the honorary secretary, W. S. Milliard told him that the bird was the Yellow Throated Sparrow. Milliard also told him about the variety of sparrows. The conversation left a deep impact on the young boy who had never thought there were so many types of birds, leave alone so many kinds of sparrows in the world. That day Salim decided that he wanted to know everything there was to learn about birds. He learnt how to identify a bird and stuff it for preservation. He was going to be an ornithologist.

Surprisingly, Salim Ali had no university degree. Although he joined college, his intense dislike of algebra and logarithm did not allow him to continue his studies. As a young man, Salim had to face years of unemployment and hardship. There were hardly any jobs available for ornithologists in India and so in 1919 Salim moved to Burma to look after the family mining and timber business. It was a rewarding experience for the naturalist as there were endless opportunities for exploring the forests of Burma. Though he failed in learning the intricacies of the business, he explored for birds in the thick jungles of Burma.

After returning to India, Salim did a course in Zoology and was appointed as a guide at the museum in Bombay. But his heart was not in it, and he tried to get a job as an ornithologist with the Zoological Survey of India but was rejected since he did not have an M.Sc or Ph.D degree.

When Salim Ali heard of an opening as a guest lecturer at the newly opened natural history section of the Prince of Wales Museum in Mumbai he decided to study further in order to qualify for the job. Salim went to Germany and trained under Professor Stresemann, an acknowledged ornithologist in Berlin.

However, when he came back to India he found out that there were still hardly any opportunity in his profession. Another man would have given up in disgust, but not Salim. Luckily, his wife had a small income that could support him. They moved to a small house at Kihim, across the harbour. It was a quiet place set amongst a thick patch of trees. When the monsoon came, Salim Ali found a colony of Weaver birds on a tree near his house. Not much was known then about these birds and this was his golden opportunity to study them. For three or four months he patiently watched their activities for hours. The publication of his findings in 1930 brought him recognition in the field of Ornithology and he won high acclaim.

He decided to create an opportunity. He went to the Bombay Natural History Society (BNHS) and offered his free services for conducting regional ornithological surveys. His only condition was that he should be provided funding for camping, and transportation costs.

The BNHS spoke to some Princely States about the idea. The States were only too eager to have their birds recorded for posterity, and they readily agreed to this novel plan. From there onwards Salim began his life as a nomad, moving from state to state and recording the variety of bird life in India. Salim was living the best years of his career. The long hours spent in the field studying birds made him one of those rare Indians who really knew each and every part of their country. Ali's wife accompanied him during these travels although the camp conditions were particularly hard for her.

Says Salim Ali "It is seldom one gets an opportunity in life to do what one wants to do. I think the best results are those when you are doing something worthwhile which you enjoy doing without the motivation of material reward. I have been exceptionally fortunate in that I was able to indulge my abiding interest of ornithology and natural history in some small measure and add to my scientific knowledge. It is something about which I can feel truly gratified."

When Salim Ali lost his job at the financially troubled museum, he decided to devote his time to the study of some weaverbirds he found on a tree near his house. Until then nothing much was known about weaverbirds. For almost three or four months he studied the birds closely and later published his findings. The publication, in 1930, brought high acclaim!

The months he spent watching the weaver birds also taught Salim Ali the importance of making first hand observations and not accept blindly the notions of anyone however famous. He checked and rechecked his observations many times and never jumped to conclusions. A famous instant of his observations was on the growth of tail feathers of the rocket-tailed drongo. A leading ornithologist claimed his observations were wrong. But eventually, Salim Ali's observations were proved right. His discovery of Finn's Baya is also a major one. This bird was supposed to be extinct for 100 years, until he discovered it in the Kumaon hills and, managed to save the 200 - year old institution from closing down due to lack of funds by writing to the then Prime Minister Nehru for help. The Prime Minister immediately came to the rescue and gave the society funds to tide over its difficult period.

It was his sincerity that won him numerous awards and medals from all over the world including the J. Paul Getty International Award, the Golden Ark of the International Union for Conservation of Nature, the golden medal of the British Ornithology Union and a Padma Shree and Padma Vibhushan from the Indian Government. Dr. Ali was nominated to the Rajya Sabha in 1985. But despite all the fame and adulation showered upon him, Dr. Ali remained what he was as a nine year old - an ever curious person with a passion for birds. Dr. Salim Ali authored numerous books, including the Handbook of the Birds of India and Pakistan (co-authored with S. Dillon Ripley) in ten volumes, books on birds of Sikkim, Kutch etc. His Book of Indian Birds with its lively descriptions and pictures can be used even by the common man. Dr. Salim Ali passed away in 1987 at the age of 91, after a prolonged battle with prostate cancer.

Salim Ali was as active in the field of conservation as he was in Ornithology. He wrote, 'For me wildlife conservation is for down to earth practical purposes. This means - as internationally accepted - for scientific, cultural, aesthetic, recreational and economic reasons. And sentimentality has little to do with it. I, therefore, consider the current trend of conservative education as given to the young on the ground of `ahimsa' alone - something akin to the preservation of holy cows - unfortunate and totally misplaced: the interest on the capital must be used, while leaving the capital itself intact. This is how I interpret wildlife conservation, and believe that future generations should enjoy the same fun with it that I had."

He was probably the only person who had travelled to all the obscure parts of the Indian Subcontinent at one time or another of his life and knew the country, its forests and its bird life intimately. His knowledge and experience were respected and his timely intervention saved the Bharatpur Bird Sanctuary, now the Keoladeo National Park, and the Silent Valley National Park. He was a non-conformist, a man who for many years walked a lonely path divergent from the mainstream of science in India. It is a tribute to his determination and genius that at the end of his life, he had a sizeable population of the conformist mainstream following him or at least appreciating and commending his more or less single-handed efforts to present the study of the Birds of his land, the ethereal spirits of the air, to his countrymen and to the world.

Goa's only bird sanctuary is situated on the western tip of the island of Chorao along river Mandovi near Panaji. It is 1.8 sq.kms. in area. A variety of local and migratory birds can be found on this island. The sanctuary is named after one of India's most prominent ornithologists, Dr. Salim Ali. From Panaji, one needs to take a bus or a cab to the Ribandar ferry wharf and then take a ferry across the Mandovi river to the island of Chorao. The sanctuary is within walking distance from the ferry wharf at Chorao. Spread over 1.78 sq. kms. and located at the western tip of the island of Chorao along river Mandovi near Panaji. It is fully covered with mangrove species. Varieties of local as well as migratory birds frequently visit there. This sanctuary can be visited any time of the year with the permission of the Chief Wild Life Warden, Forest Department, Junta House, Panaji-Goa. The sanctuary is approachable by walk after crossing over by ferry from Ribandar to Chorao.

For nearly fifty years 'The Book of Indian Birds' has been a close companion of the recently inducted bird-watching enthusiast as well as of the seasoned ornithologist in India. It had lively descriptions and colourful pictures of all species. It made spotting a bird easy for a layman. In 1948 he began an ambitious project in collaboration with S. Dillon Repley, an ornithologist of international repute, to bring out ten volumes of 'Handbook of the birds of India and Pakistan'. This work contains all that is to be known of the birds of the subcontinent, their appearance, where they are generally found, their breeding habits, migration and what remains to be studied about them. Salim Ali travelled all over the country on this bird watching surveys. It is claimed that there is hardly a place in the country where his heavy rubber shoes have not left their mark.

The Bombay Natural History Society (BNHS) has brought out a special issue dedicated to the birth centenary of Dr Salim Ali. This issue of Hornbill-a beautifully printed coloured publication can also be marked as the second tribute to Salim Ali-the first issue being published on his 80th birthday. A total of 11 articles all recounting the golden days with Salim Ali feature in the present issue. The authors have narrated warmly and very fondly their memories with the grand old man of Indian ornithology. All of them without fail have paid tribute to the physical stamina of a frail old man, his keen sense of observation, his meticulous notings and unending questionings.

The gradual change of a classical ornithologist into an ardent conservationist, albeit with a sense of apparent skepticism, becomes real in Dilnavaz Variava's, Memories of Golden Days. S.Dillon Ripley, the lifelong friend and co-author of Salim Ali's monumental treatises, candidly mentioned "At the start of our acquaintance, conservation was not our major concern.... Later in our career we became increasingly motivated by the overriding importance of conservation". We can all recall the result of such motivation in saving Silent Valley in the early eighties.

Salim Ali, the person, has an unparalleled sense of sincerity, dedication, meticulousness and austerity have been repeatedly mentioned as his hallmark traits, few perhaps know about his "highly inflammable temper, impatience, ruthless discipline and devastating sarcasm, intended mostly to punctuate inflated egos. That such a strict disciplinarian could be taken easily with sweet and well chosen words may surprise many. That's but normal for a human.

All geniuses have contradictions which make them human. "Salim Ali was cheery and full of humour and fond of good literature," writes his friend Dillon Ripley as also his close friend Loke Wan Tho, the expatriate Chinese from Singapore of whom Salim Ali wrote about, Salim Ali as the man who had an eye for beauty. Zafar Futehally recounts how both (Salim Ali and Wan Tho) after a gruelling day's work would sit around "reading loudly from some anthology or other. Macaulay's Plays of Ancient Rome was a particular favourite" (Zafar Futehally, In camp with Salim Ali)

So we have a complete man personified in the bigger than life size image; a man who became a legend in his lifetime, a man who dared to become a non-conformist in an orthodox family, a man who never knew how to rest; and also a man who almost in a childlike manner would whisper a request for more ice-cream when dinner plates were being removed.

Salim Ali, the ornithologist par excellence emerges sharply through these pages of the personal reminiscences of his associates and friends. Apart from his autobiographical book 'The Fall of a Sparrow' not much has been written by others on the life of Salim Ali. It is wonderful how immediate a connection people make between birds and Dr. Salim Ali, even in remote and unlikely places.

✥ ✥ ✥